The Planetary Equatorium
of
Jamshīd Ghiyāth al-Dīn al-Kāshī

Princeton Oriental Studies

Volume 18

THE PLANETARY
EQUATORIUM

OF JAMSHĪD GHIYĀTH AL-DĪN AL-KĀSHĪ
(d. 1429)

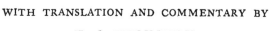

An Edition of the Anonymous Persian
Manuscript 75[44b] in the Garrett Collection
at Princeton University
Being a Description of Two Computing Instruments

The Plate of Heavens

and the Plate of Conjunctions

WITH TRANSLATION AND COMMENTARY BY
E. S. KENNEDY

1960

PRINCETON UNIVERSITY PRESS
PRINCETON, NEW JERSEY

Printed in the United States of America
By the Meriden Gravure Company, Meriden, Connecticut

To
my foster mother
Annie E. Kennedy

P R E F A C E

This book presents, with translation and commentary, the
text of a Persian manuscript in the Garrett Collection at
Princeton University. The manuscript describes the construction and use of two astronomical computing instruments invented
by the fifteenth century Iranian scientist Jamshīd ibn Masʿūd
ibn Mahmūd, Ghiyāth al-Dīn al-Kāshī (or al-Kāshānī). Our fund
of knowledge concerning the work of this individual has been
materially increased in recent years, particularly by the
studies of the late Paul Luckey in Germany, and by the fruitful
collaboration of B.A. Rosenfeld and A.P. Yushkevich in Russia.
The reader will find their publications listed in the bibliography at the end of this volume.

Little enough has been known about the life of Kāshī (as
we shall call him hereafter), and we have been able to supplement
that little, principally by examining the introductions and
colophons of Kāshī manuscripts. The results are assembled in
the biographical sketch immediately following the table of
contents.

The succeeding section recites the relations between our
anonymous Persian text and the two versions of a book by Kāshī
himself on which it is based.

There follows the facsimile text and translation, arranged
for immediate reference with corresponding pages and lines
opposite each other. After this comes the commentary to the
text. In general, topics are treated in the order in which
they appear in the manuscript. Where it has been found convenient to digress from this order, a statement has been inserted
indicating the proper section. When sections are related to a
particular passage in the text, this passage has been specified
in parentheses following the section title. All such references
to the text give folio and line, separated by a colon.

It is assumed that the reader is familiar with the leading
concepts of Ptolemaic astronomy, of which [41] and [42] contain

readily available expositions. Here and in the sequel,
numbers enclosed in square brackets refer to items in the
bibliography at the end of the book.

The bulk of the manuscript is taken up with an instrument called by Kāshī Ṭabaq al-Manāṭiq, which we translate as
"Plate (or Tray) of Heavens." It is an example of the class
of devices known as equatoria, analogue computers on which
the various Ptolemaic planetary configurations were laid
out to scale. They thus yield solutions to such problems
as that of finding the true longitude of a planet at a given
time. Kāshī's equatorium is only one element in a tradition
stretching through sixteen centuries in time, and in space
from Western Europe to Central Asia via North Africa and the
Near East. The story of the other instruments, and the Plate
of Heavens' place therein, has been delineated admirably by
D.J. Price in his publication [42] of the Chaucerian equatorium - there is no need to repeat it here.

The second instrument, the "Plate of Conjunctions"
(Lawh al-Ittiṣalāt), is a simple device for performing a
linear interpolation.

A number of the problems raised by the manuscript have
been dealt with in a preliminary way in papers published
over a period of years. This is an attempt simultaneously
to dispose of the remaining problems, to revise, complete,
and correct previous solutions, and to put before the
public the primary source on which they are based. In
addition to presenting the philologians with an opportu-
nity to pick apart the work of a translator whose formal
training has been in mathematics, this will serve to
present intact a medieval scientific work, a very small
link in the chain which leads from the tallystick to the
electronic computer. For the manifold shortcomings of the
result the editor, like the anonymous author of the manus-

P R E F A C E

cript, craves correction "with the musk-dripping pen" of
forebearance.

Acknowledgments

In thus bringing to a close a task, the working out
of which has been a source of much satisfaction over a long
time, it is pleasant indeed to set down the names of some
of the many friends who have been helpful. Professor Yahya
Armajani called to my attention the equatorium manuscript
in the first place. In the office of the late Professor
George Sarton I was privileged to associate with Dr. Alexander
Pogo, and from him gained my first real insight into planetary motions. Professors N. Seifpur Fatemi and Jalal Homa'i,
and Dr. Iraj Dehqan answered many questions connected with
the reading of the Persian text. The commentary has been
improved as a result of many discussions with Professor Derek
Price, whose book on the Chaucer equatorium will continue
to maintain itself as the model of publications of this
genre.

The contributions acknowledged above, important as they
are, can be regarded as peripheral and as not involving the
individuals in errors which the book may contain. It remains
to name another associate, one who cannot escape his portion
of responsibility for whatever merit this work has. To the
counsel and example of O. Neugebauer, best of friends and
keenest of critics, I owe such competence as I have acquired
in the history of the exact sciences.

The work was made possible by the Rockefeller Foundation,
Brown University, the Institute for Advanced Study and a
Fulbright grant. The Princeton University Department of
Oriental Languages and Literatures and the Princeton Univer-

P R E F A C E

sity Press have been uniformly cooperative in making arrangements for publication.

Copy for the photo-offset reproduction was typed by
Mrs. Kawthar A. Shomar with her customary accuracy, elegance,
and dispatch.

E.S.K.

C O N T E N T S

xi

C O N T E N T S

CONTENTS

COMMENTARY

C O N T E N T S

CONTENTS

LIST OF FIGURES
(In the text)

(Others)

C O N T E N T S

Biographical Material on the Inventor

Our earliest fixed point in Kāshī's life is 2 June, 1406
(12 Dhū al-Hijjah, 808). On this date he observed in Kāshān,
his home town in central Iran, the first of a series of three
lunar eclipses (see [23]*, f.4r.)

He is the author of an Arabic treatise on the sizes and
distances of the heavenly bodies, called Sullam al-samā' [22].
Most of the extant copies of this work are undated, but in
[34], p.50, Krause reports that one of the Istanbul copies
claims the book was finished on 1 March, 1407 (21 Ramaḍān,
809). This is confirmed by Ṭabāṭabā'ī ([52], p.23) on the
basis of a copy at the Shrine Library in Meshed, Iran. Since
this date falls during the run of eclipse observations, it
follows that the treatise must have been written in Kāshān.
It is dedicated to a certain wazīr designated in the manuscript
only as Kamāl al-Dīn Maḥmūd. Search through general histories
of the period has failed thus far to produce any information
whatsoever as to the identity, jurisdiction, or political
affiliation of this individual. A certain Maulā Kamāl al-Dīn
Maḥmūd al-Sāghirjī mentioned by Khwāndamīr ([33], vol.3, p.513)
fulfills the requirements as to name and rank, but cannot have
been contemporary with Kāshī. It is probably to this individual
that Ṭabāṭabā'ī ([52], p.23) refers.

Sometime during the year 816 A.H. (1413/14) Kāshī completed
the Khāqānī Zīj [23], written in Persian and the first of his
two major works. In the introduction to it he complains that
he had been working on astronomical problems for a long period,
living in penury in the cities of ʿIrāq (doubtless ʿIrāq-i
ʿAjamī, Persian ʿIrāq), and most of the time in Kāshān. Having
undertaken the composition of a zīj, he would have been unable

---

* Numbers in square brackets refer to items in the
bibliography.

to complete it except for the timely beneficence of the prince
Ulugh Beg, he says, and to him he dedicated the finished work.
The fact that the longitude of Shīrāz was taken as base for the
mean motion tables (see Section 18 in the commentary below) does
not prove that he worked there part of the time, although he may
have done so. The place had been a center of astronomical
activity on and off for many centuries, and Kāshī may simply
have chosen as base a location better known than Kāshān. From
a remark in his Miftāḥ ([47], p.176) it is clear that he was in
the neighboring city of Isfahān at one time or another.

Next comes a very short (seven page) Persian treatise [20]
on astronomical instruments written in January 1416 (Dhū al-Qaʿda,
818) and dedicated to a Sultan Iskandar.

Soon after this, on 10 February, 1416 (10 Dhū al-Hijjah,
818), and in Kāshān, the first version of the Nuzhat al-Hadāʾiq
was completed. The book is Kāshī's own description of the
equatorium he invented. It names no patron. When next heard
from, Kāshī has joined the group of scientists at the Samarqand
court of Ulugh Beg. There he stayed for the rest of his life.

His career commenced during the long reign of Tamerlane.
When the latter died in 1405 he was eventually succeeded by his
son Shāhrukh, whose rule outlasted Kāshī's life span, and who
throughout evidently retained some sort of hegemony over all of
Iran. During this time Shāhrukh's son Ulugh Beg was the ruler
of Samarqand, and eventually as head of the Timurid dynasty
survived his father for a short time.

The Iskandar referred to above can hardly have been any
other than the son of Qara Yūsuf ([12], p.127) second ruler of
the Black Sheep Turcoman dynasty which established itself mainly
in Azarbaijan and Mesopotamia, encroaching eventually into Fārs.
Iskandar was twice defeated by Shāhrukh. Like Ulugh Beg, he
achieved primacy in his dynasty only long after the death of
Kāshī.

2

INTRODUCTION

The curious spectacle of a scientist dedicating successive
writings to minor potentates in rival dynasties invites specu-
lation. It is tempting to liken his actions to those of a
modern scholar alternately wooing one and another of the affluent
learned foundations in hopes of an ever more princely grant.
Kāshī's short treatise on astronomical instruments [20] contains
little beyond what must have been common knowledge to any competent astronomer of the day, and its composition can have cost
him only the time required to write it down. He may have turned
it out and dedicated it to Iskandar in order to counteract the
effect of the earlier dedication to a Timurid. But ignorance
of the details of political vicissitudes makes further conjec-
tures unprofitable.

In 1417 Ulugh Beg commenced building in Samarqand a madrasah
([16], p.54), a school to house students of theology and the
sciences. This impressive tiled structure is still admired by
tourists from all over the world. Shortly after its completion
the construction of the observatory was begun.

It was during these operations that Kāshī arrived; we do
not know exactly when. Several sources (e.g. [33], vol.4, p.21)
have him accompanied by a fellow-townsman and astronomer, Muʿīn
al-Dīn al-Kāshī, according to Ṭabātabāʾī ([52], p.6), a nephew.
From this period also is a document of the greatest interest,
a letter from Kāshī to his father in Kāshān. It has been published (as [53]) in the original Persian, collated and annotated
by M. Ṭabātabāʾī from two manuscripts, one of which is in the
library of the Madrasah Sepahsālār in Tehrān. The letter merits
publication in facsimile with a translation into some European
language. Pending this we outline its contents below.

The epistle begins with a quotation from the Qurʾān indi-
cative of filial piety, and a statement that the writer has been
too busy with the observatory to do anything else. He writes
that the sultan is an extremely well educated man, in the Qurʾān,

3

in Arabic grammar, in logic, and the mathematical sciences.
As an illustration of the latter he tells how the king, while
on horseback, once computed in his head a solar position
correct to minutes of arc. Kāshī then goes on to describe how,
upon his arrival at Samarqand, he was put through his paces by
the sixty or seventy other mathematicians and astronomers in
attendance there. He gives as examples four of the problems
propounded to him. The first involved a method of determining
the projections of 1022 fixed stars on the rete of an astrolabe
one cubit in diameter. The second required the laying out of
the hour lines on an oblique wall for the shadow cast by a
certain gnomon. The third problem demanded the construction
of a hole in a wall, of such a nature that it would admit the
sun's light at, and only at, the time of evening prayer, the
time to be that determined by the rule of Abū Hanīfah. Lastly,
he was asked to find the radius, in degrees of arc on the
earth's surface, of the true horizon of a man whose height is
three and a half cubits. All these and others, says Kāshī,
which had baffled the best minds of the entourage, he solved
with ease, thus quickly gaining intellectual paramountcy among
them.

He held a low opinion of the rest of the sultan's scientific staff in general, in spite of the fact that Ulugh Beg's
astronomical bent had stimulated the study of mathematics in
Samarqand for the past ten years. The only one who gave him
any competition at all was Qādīzādah al-Rūmī ([51], p.174),
and Kāshī recounts in detail two occasions on which he worsted
this individual. One of these was brought on by Qādīzādah,
who had been expounding the famous zīj of al-Bīrunī, the
Masudic Canon [5], when he ran into difficulties with the
proof of a theorem. Using the immemorial gambit of any mathematics teacher when faced with a like situation, he told the
class, which included Ulugh Beg himself, that there must be a

fault in the text, he had best compare it with a better copy
elsewhere. After two days he was still stuck at the same place,
when Kāshī just happened to turn up, explained the proof off-
hand, and showed that the manuscript was correct.

In spite of this, and other exhibitions of tact and
erudition on the part of Kāshī, he assures his father that
relations between the two of them are of the most amicable.
He agrees that Qāḍīzādah is the only one of the lot who knows
much about the Almagest, although he is deficient in observational technique. The views of Qāḍīzādah on the matter have
not come down to us, but it may or may not be of significance
that in the prolegomena to Ulugh Beg's zīj ([54], p.5) Qāḍīzādah is the first of the two to receive honorable mention and
extravagant praise.

Kāshī uses up a good deal of space telling his father
about the ignorance of a certain Badr al-Dīn. This man, he
writes, having been through a few propositions of Euclid which
he is unable to apply, is like one who knows several rules of
(Arabic) grammar but can write no Arabic. He states further
that Badr al-Dīn is a liar. Kāshī's motive in reporting on
this person is not clear.

The letter describes the status of the work in progress
on the construction of instruments for the observatory. It
closes with a detailed explanation intended to make clear to
the layman why the taking of observations for a complete set
of planetary parameters is a long process, and cannot be
completed in as short a time as a year, or anything like it.

In the middle of Sha'bān, 827 (July, 1424), Kāshī finished
his unprecedentedly precise π-determination, al-Risālat al-
muhiṭīyah. Written in Arabic, this masterpiece of computatio-
nal technique has been published in German [36] and Russian
[47] editions, both excellent. It has no dedicatory passage.

From the same month two lunar years later (June, 1426)

dates the Samarqand rescension of the Nuzhah, and on 3 Jumādī
I, 830 (2 March, 1427), Kāshī completed his second major work,
the Miftāḥ. It is dedicated to Ulugh Beg, and has recently
been published [47] with Arabic text, Russian translation, and
commentary, by Rosenfeld and Yushkevich. Studies based on it
are [35], [37], [44], and [7]. In the introduction to the
Miftāḥ Kāshī gives a partial list of his own works. Other
than titles already mentioned above, the following appear:

Risālat al-watar w'al-jaib, also known as Risālah fī
istikhrāj jaib darajat wāhidah, is apparently extant both in
a lithographed edition printed in Tehran in 1889, and in manuscript ([51], p.174). Marginal notes in [23], f.32r, and the
lithograph edition of [17], p.3, state that the treatise was
incomplete when Kāshī died and that it was finished by Qādīzādah. A copy of the lithographed collection of which it is a
part is in the Parliament (Majlis) Library in Tehran. It
describes an elegant iterative method of computing the sine
of one degree to any required accuracy. No translation of
this risālah has been published, although a commentary on it
is available in French [54] and in Russian [47] translations,
and a considerable literature in European languages has accu-
mulated about it (see, e.g. [1]).

The Zīj al-tashīlāt is not extant as such, but is probably the set of tables and accompanying explanation for a
simplified method of computing planetary positions as worked
out by Kāshī. In his Khāqānī Zīj ([23], ff.142r-155r) these
tables occupy a section distinct from the planetary tables of
standard type.

In addition to those listed in the Miftāḥ al-hisāb, the
following treatises were written by or attributed to him:

Miftāḥ al-asbāb fī ʿilm al-zīj, listed in the Mosul catalogue [40].

Risālah dar sākht-i asturlāb listed in the Meshed catalogue

([38], Ms. math. 84).

Risālah fī maⁿrifat samt al-qiblah min dā'irat hindiyah, also listed in the Meshed catalogue.

Risālat ⁿamal al-darb bi'l-takht w'al-turāb is also included in the Tehran lithographed edition of 1889 of which a copy is in the Parliament Library in Tehran.

Al-risālat al-iqlīlāminah is mentioned by Kāshī himself in the Samarqand version of the Nuzhah ([19], p.311).

Kāshī's statements about the length of time required to complete the observations proved all too true. In the prolegomena to the zīj ([54], p.5) based on them, his royal patron Ulugh Beg, laments his death early in the course of the work. His collaborator and rival, Qādīzādah, also died before the zīj was finished. On the title page, the India Office copy of the Khāqānī Zīj [23] bears a note saying that on the morning of Wednesday, 19 Ramadān, 832 (22 June, 1429), at the observatory outside Samarqand there died the "mighty master, Ghiyāth al-Millah w'al-Dīn, Jamshīd." According to Tabātabā'ī ([52], p.19) the incomplete copy of the same zīj in the Shrine Library at Meshed has the same annotation. That Kāshī left behind him more than his scientific works is witnessed by the British Museum manuscript of his Miftāh ([6], p.199) which has the following curious colophon:

> Verily I finished copying this honorable
> manuscript on the second day of the month of
> Shawwāl, year 997, (14 August, 1589). It was
> copied (or checked ?) by the one (who is)
> indigent (for the sake) of God, al-Razzāq,
> son of ⁿAbdallāh, son of ⁿAbd al-Razzāq, son
> of Jamshīd, son of Masⁿud, son of Jamshīd,
> the author of this noble book.

We lack evidence on which to pass judgment, or even to assess, Kāshī's personal character. Concerning him the author of the

THE PLANETARY EQUATORIUM

Haft iqlīm (see [45], vol.ii, p.45) has this to say:

> The former (Kāshī) was ignorant of
> the etiquette of courts, but Ulugh Beg was
> obliged to put up with his boorish manners
> because he could not dispense with his
> assistance.

In the letter to his father he does not depict himself as
the proverbially shrinking violet. At the same time this was
a personal communication addressed to a parent, and presumably
not intended for publication. And on the basis of the evidence
at hand we can only agree with his own estimate - he was the
best of the lot at Samarqand.

In the closing appendix to the revised Nuzhah, "On the
Naming of the Instrument," ([19], p.312) Kāshī whimsically
relates a suggestion from some of his friends, that his equatorium be called Jām-i Jamshīd,

> ... Jamshyd's Sev'n-ring'd Cup, where no one knows,

a likening of the instrument with its seven planetary deferents
to the magic divining goblet of the mythical Iranian king, dis-
coverer of the uses of wine, and whose name Kāshī bore.

> They say the Lion and the Lizard keep
> The courts where Jamshyd gloried and drank deep.

As for his scientific attainments and his place in the
history of science, here we can operate from much firmer ground.
He was first and foremost a master computer of extraordinary
ability, witness his facile use of pure sexagesimals, his invention of decimal fractions (cf. [44] and [37], p.102), his wide
application of iterative algorisms, and his sure touch in so
laying out a computation that he controlled the maximum error
and maintained a running check at all stages.

His equatorium marked the most extensive development ever
given to this class of instrument. In particular his was the
only mechanical device with which a determination of the planetary

latitudes was possible. If we retain mental reservations as to
the practicality of the results, there can be no doubt of the
ingenuity of the descriptive geometry involved.

He seems to have been a completely competent observer and
astronomical technician, neither ahead of nor behind his times.
The same statement can be made of his work in planetary theory.
He accepted unreservedly the notion, not contained in the Almagest, that the moon, inner planets, sun, and other planets move
in contiguous bands about the fixed earth, and that hence it
was possible to compute in terrestrial units the mean distance
of, say, Saturn from the earth. His contemporaries were therefore overgenerous in calling him "the second Ptolemy" ([33],
vol.iv, p.21), but the next generation was equally sanguine in
calling a mathematician of their own time "the second Ghiyāth
al-Dīn Jamshīd" ([53], p.60).

The History of the Text and Its Versions

The Persian manuscript published in this volume is an
apparently unique copy of a tract composed by some individual
now unknown, between the years 1481 and 1512, and probably in
Constantinople. This time spans the reign of the Ottoman
Sultan Bayazid II, to whom the book is dedicated. Constantinople is mentioned in the text as this ruler's capital, and
its longitude is taken as base for the planetary mean motion
tables.

If the author is unknown, his prime source is not. He
specifically states that the astronomical instruments he
describes were invented by Jamshīd (al-Kāshī), and his work
can be characterized as largely a translation into Persian
of selected parts from Kāshī's own Arabic description of the
same two instruments.

The latter book is called <u>Nuzhat al-Hadā'iq</u> (A Fruit-

9

Garden Stroll) and is extant in two versions. The first was
written in Kāshān and completed on 10 February, 1416. The only
copy known to have survived is Number 210 in [48], a microfilm
of which has been made available by the officials of the India
Office. The copy is modern, having been completed on 2 December,
1863. It is written in an easily legible nastaᶜlīq hand, but
very carelessly, and all the figures are missing. This Kāshān
version we refer to as NK in the sequel.

In June, 1426, just three years before he died, Kāshī
completed a rescension of the Nuzhah. This was after he had
moved to Samarqand to work at Ulugh Beg's observatory there.
Following the colophon of the original material, which was
changed only in minor details, he added a set of ten appendices.
These describe additional techniques for utilizing the instruments, and improvements or changes in the construction of their
parts. It will be shown in the commentary (Section 42) to our
text that the author made use of the Samarqand rescension, which
we will abbreviate as NS. No manuscript copy is known to exist,
but in 1889 it was printed in the Tehran lithograph edition of
several of Kāshī's works. The example in the translator's pos-
session is bound and paginated (pp.250-313) with the Miftāh
al-Hisāb. It is written in a fair naskh hand, very carefully,
and with text corrections in the margin. The space on page 261
for the main figure, however, has been left blank.

A table of contents of NS and NK follows, combined with
a concordance of corresponding sections in our text.

10

INTRODUCTION

Table of Contents of the Nuzhat al-Hadā'iq
With a Concordance of the Kāshān (NK),
Samarqand (NS, Tehrān lithograph),
and Persian Versions

11

INTRODUCTION

T E X T
and
T R A N S L A T I O N

    The numbers along the left edge of each page of the
translation identify corresponding lines of the text facsimile
on the opposite page. Restorations to the text are indicated
by the usual square brackets; in general, words enclosed in
parentheses have no equivalents in the text, but have been
added to clarify the meaning of the relevant passage.

    Marginal additions to the text are of two kinds, scribal
omissions from our copy, and explanatory notes. The former
were written in, perhaps at the time the copy was made,
presumably by checking it against another copy. These have
been inserted into the body of the translation, set off by
asterisks at the beginning and end of the insertion. The
marginal notes proper, however, have been translated as such
and appear in the margins of the translation.

رساله ٤ العمل باسهدآية من قبل النجوم

A Short Work on the Operation
of the Easiest Instrument
Having to Do with the Planets[*]

---

[*]This title, in Arabic and written in a hand
different from that of the rest of the manuscript,
need not be regarded as the original title of the
work.

شکر و سپاس و ستایش بی قیاس لایق حضرت

عزت ... ربانی باشد که اطباق سموات را بذر...

غیر ... ثواقب و جواهر زواهر

ثوابت مرصع گردانید چنانکه

سفلی ... کاملداش حدوث مکونات سفلی را

مرتبط با وضع علوی ساخت قادری که بدست

قدرت و تقدیر چندین اشخاص منیر را در

قالب تصویر آورد لا اله الا الخالق والامر فتبارک

الله احسن الخالقین و صلاة صلوات و

تحف تحیات برقطب فلک رسالت و عدا

TRANSLATION

f.2v

1   TO THE PRAISE OF GOD

2   Thanks and [praise][1] and incomparable glory
      is the due of the

3   Power,[the Creator][2] who has inlaid the layers
      of the heavens with

4   glistening pearls, with shooting [stars][3], and
      glittering jewels,

5   the planets and the fixed (stars). The Sage
      who,

6   impelled by His complete [wisdom][4], wrought
      the bringing into existence of the inferior
      (planets)

7   together with the situations of the superior
      (planets); the (Al)mighty, who by the hand of

8   power and destiny moulded how many shining
      bodies.

9   "Verily his are the creation and the command;
      blessed be

10   God, the best of Creators"[5], and benediction
      and praises and

11   the gift of salutation upon the Pole of the
      Heaven of Prophecy and the Circlet of the

---

1. Wormeaten, read سپاس

2.   "   ,   "   صانعى

3.   "   ,   "   بكواكب

4.   "   ,   "   حكمت

5. Qur'ān, 7:52 (ed. of Flugel).

19

آسمان جلالت محمد مصطفی و بر آل و اصحاب او

که سود سپهر استاد و سبهر بکوم راقند الله باد و بعد

پوشیده نماند که شریفترین نوعی از انواع علوم ریاضی

علم نجوم است که فضل انسانی را بر او اقتضا آن تشرف

اطلاع بر کیفیات حرکات کواکب و طول و عرض

و کیفیات اوضاع ایشان و مقامات هر یک

در بروج و نظامات و معرفت و قات صلوة

و سمت قبله و سایر بلاد و ابعاد اجرام هر یک

از کواکب از زمین و مساحت میان بلاد حاصل

می‌شود و حصول جمله این مقاصد میسر نگردد از بی

و معرفت آن موقوفست بحساب از تفریق و

تضعیف و جمع و تنصیف و ضرب و قسمت و

اعمال بعضی از تعدیلات و صعوبت آن ظاهر

f.3r

1   Sky of Glory, Muḥammad the Chosen (One) and
upon his descendants and his companions,

2   who are the beneficent (planets) in the sky
of leadership and the sphere of the stars of
imitation. However,

3   let it not remain covered that the most honorable branch of the branches of the science
of mathematics

4   is astronomy, which supplies the human soul
with the acquiring of

5   that honor (i.e. astronomy), knowledge of the
amount of the planetary motions in longitude
and latitude,

6   and the modes of their conditions, and the
stations of each one

7   in the zodiacal signs, and the (planetary)
sectors, and the knowledge of the times of
prayer,

8   and the direction of Mecca, and the other
cities, and the distances of the bodies of
each

9   of the planets from the earth, and the distance between cities.

10   But the attaining of all these objectives is
impossible except by (means of) an astronomical handbook (a z̄īj),

11   and the knowledge of it is dependent upon
computations: on subtraction, and

12   halving, and addition, and doubling, and
multiplication, and division, and

13   operations with various (astronomical) equations, and its difficulty is apparent

21

پدیدآوردن سخن سودیاست واستادِ مُحَقَّق و مُحَرَّر
مُدَقَّق مُکَمِّل علومِ اوایل و کاشف مُعضَلاتِ مسائل
افتخار الحکماء فی العالم مولانا اعظم مولانا غیاث
الملة والدین جمشید برد الله مضجعه آلّیّه تصدُّر کرد و
وآنرا طبقُ المناطق نام نهاد و درسالۀ دیگر کیفیت
صنعتِ علم آن پرداخته که بدان اُلا معمول
رایج باسهل طریقی و اقلِّ زمان عمل کرد و پی شود
وجدان احتیاج بهرب و قسمتِ حساب
ندارد و الی زماننا هذا اسهج احد از آحادِ علما
باوجودِ آنکه هیچکس همّت بر اجتماعِ آن فضایل و
افضالِ مقصوره داشته اند نقاب حجاب از
چهرۀ آن مخدَّرتا بر داشته اند و ازین بی حدّ
مخزونِ بعالمِ روز زیاورد و وبین فقیر بتوفیق

f.3v

1   and known and clear and manifest, and so the
    meticulous master, the investigator, the

2   scrutinizing author, the completer of the
    primary sciences and the solver of difficult
    questions,

3   the pride of the savants of the world, our
    most mighty master, our master Ghiyāth

4   al-Millah w'al-Dīn (the Refuge of the Congre-
    gation and of the Faith) Jamshīd, may God
    cool his resting-place, invented an instrument

5   and called it the Plate of Heavens, and
    prepared a treatise describing its

6   construction and operation, such that with
    this instrument the problems commonly solved
    with a

7   zīj may be solved by the easiest methods and
    in the shortest time,

8   and it does not require much multiplication
    and division and computation.

9   And up to this our time, no single one of the
    scholars,

10 notwithstanding all the effort toward the
    carrying forward of learned and excellent
    illustrious deeds,

11 they were deficient. The veil-curtain from
    before

12 the face of that virgin maiden they did not
    raise, and from behind the curtain of

13 existence to the world she was not made
    manifest. But this destitute one, by grace of

ارادت بجون و یاس دولت روز افزون

پادشاه جهان، فرمان فرمای دوران سلطان

آسمان سریر خورشید ضمیر ستاره لشکر مشتری نظر

ظل الله فی الارضین قهرمان الماء والطین

قامع الکفرة والمشرکین بیت

آنت ناید جق صورت امن وامان

نص کتاب ظفر مهدی آخر مان

السلطان بن السلطان ابوالفتح سلطان بایزید

بن سلطان محمد خان خلد الله ملکه وسلطانه و ابد علی

علی العالمین برّه واحسانه این آنت را ساخت

و تقویم کواکب و عروض ایشان و خسوف

و کسوف عمل کرده با زیج موافق کردانید

و در آنکه عمل کردند بود عجالة الوقت را اورق

TRANSLATION

f.4r

1   the ineffable will and bounty of the daily-
    increasing government of the

2   Ruler of the World, the Issuer of Commands of
    the Ages, the Sultān

3   of the Heavenly Throne, the Sun of Conscience,
    the Star of the Jupiter-appearing Army,

4   the Shadow of God on both Worlds, the Champion
    on Sea and Land,

5   the Subduer of the Infidels and Idolaters. A
    Couplet:

6   The symbol of (what is) supported by
        Right (i.e., God), the aspect of peace
        and security,

7   (He is) the very letter of the book of
        victory, the Mahdī and the Last of (this)
        Age,

8   the Sultān, son of Sultān Abū al-Fath, Sultān
    Bāyazīd,

9   son of Sultān Muhammad Khān, may God perpe-
    tuate his dominion and his kingdom, and may

10 his beneficence and his goodness endure upon
    both worlds, (the author) constructed this
    instrument,

11 and operated (with it to find) the true longi-
    tudes of the planets, and their latitudes, and
    lunar

12 and solar eclipses. (The results) agreed with
    (those obtained by use of) a zīj.

13 And of whatever has been done at present (i.e.,
    this work itself) having been

چند نوشته مهرِ دِوِض سدَّهَ سینه و عتبهٔ علیّه
کشت ٭ فان وقع فی جهۀ القبول فهو غایة
المأمول و نهایة المسؤل ملتمس انکه چون برسهوی
و خطایی اطلاع یابند بقلم سکبار ٭ و خانهٔ که
نگار اصلاح فرمایند و ترتیب این رساله برو
مقاله لایق افتاد مقاله اوّل
در صنعت جنس المناطق مشتمل بر پنج باب و تمام
باب اوّل در صنعت قرص و خلقت
دوم در رسم اوجات و مراکز
و مناطق و نقط محاذات و معدل المسیر باب
سیم در رسم قطر استوا و نقطه عرض و خطوط
آن باب چهارم در صنعت عضاد
و مسطره و تقسیم آن باب پنجم در رسم

f.4v

1   written on a few sheets, it is presented to
the mighty and glorious (one), the Sublime
Porte.

2   And if acceptance be possible it (would be)
the extreme of

3   that which is hoped for, the utmost of that
which is requested. It is requested that if
information of a slip

4   or a mistake is found (it) be corrected with
the musk-dripping pen and the reed (pen which
is) the pearl-portrayer.

5   And the arrangement of this disquisition came
to be worthy of two

6   treatises. (Contents of) THE FIRST TREATISE:

7   On the Construction of the Plate of Heavens,
containing five chapters and a conclusion.

8   Chapter One, On the Construction of the Disk
and the Ring.

9   Chapter Two, On the Drawing of the Apogees, and
the Centers,

10   and the Deferents, and the Opposite Point, and
the Equant. Chapter

11   Three, On the Drawing of the Equating Diameter,
and the Latitude-Point, and their Lines.

جداول اوساط و غیرها خاتمه در

صفت لوح اتصالات مقاله دوم

در علم آلت مشتمل بر یازده باب

اول در ترتیب آلت باب دوم در استخراج

اوساط کواکب باب سیوم در استخراج

تقویم آفتاب باب چهارم در استخراج

تقویم قمر باب پنجم در استخراج

تقویم کواکب متحیره باب ششم در

معرفت بعد مقامات باب هفتم در

معرفت عروض کواکب باب هشتم در

معرفت ابعاد کواکب از مرکز عالم باب نهم

در معرفت رجعت و استقامت باب دهم

در معرفت مطالعات و حجی و بُعد وتری باب

TRANSLATION

یازدهم در معرفت خسوف باب دوازدهم
در معرفت کسوف باب سیزدهم در توقیت
وسط تحویل باب چهاردهم در معرفت
ارتفاعی حقیقی و مرئی و معرفت ارتفاع مرئی یے
از حقیقی و اختلاف منظر از این ارتفاع باب
پانزدهم در معرفت تعدیل الایام بلیلها خاتمة
در عمل بلوح اتصالات مقالهٔ اول
در صنعت آلت موسوم نطق المناطق مشتمل برپنج
باب دخاتمه باب اول در صنعت قوس
و علقه چون خواستند که این آلت را بسازند قوصی
صحیح الاستداره بسازند از نحاس یا صفر
یا شبه یا خشب صلب بخم صفیحهٔ اسطرلاب
هرچند بزرگتر باشد بهتر بود دائل قطر او یک

T R A N S L A T I O N

f.5v

1   Eleven, On the Determination of Lunar Eclip-
    ses. Chapter Twelve,

2   On the Determination of Solar Eclipses. Chap-
    ter Thirteen, On the Determination of

3   the Mean of Transfer. Chapter Fourteen, On
    the Determination of the

4   True [from]*the Apparent Altitude, and the
    Determination of the Apparent Altitude

5   from the True, and Parallax in the Altitude
    Circle. Chapter

6   Fifteen, On the Determination of the Equation
    of Time. Conclusion,

7   On the Operation of the Plate of Conjunctions.
          THE FIRST TREATISE

8   On the Construction of the Instrument Called
    the Plate of Heavens, containing five

9   chapters and a conclusion. CHAPTER ONE. On
    the Construction of the Disk

10 and the Ring. If it is desired to make this
    instrument, make a disk,

11 truly circular, of copper, or brass,

12 or yellow copper, or hard wood, like the plate
    of an astrolabe,

13 the bigger the better. Its diameter should be
    at least one

---

*For ١و read زا.

ذراع بود و اگر دو یک ذراع بود بکستر باشد چه
عمل صحیح می‌آید و بر وحلقه ترکیب کنند چون حجره
اصطرلاب چنانکه صفحه در حرکت کند و از وی معو
کنزد و سطح حلقه با سطح صفحه چون یک سطح مستوی
کنزد و بدان طریقه که محیط صفحه زبانه باریک
بسازند و در تمام مقعر حلقه حفری کنند تا آن زبانه
در و در آید و استوا آید او را بسطر و شا قول متحان
کنند چنانکه معلوم صناع است پس مرکز قرص
بیخ دایر بر حلقه رسم کنند و ایره اول را بدو از دو
قسمت راست کنند تا دو آیر بنجا نه بدو نگهم
شود و در ما بین اول و ثانی اسامی بروج بویید
و توالی از جانب یسار اعتبار کنند چون خود را
بر مرکز قرص فرص کرده باشند و ثانی را به هنر

f.6r

1   cubit, and if it be two or three cubits,
2   operating with it will be more accurate. And
     mount around it a ring, like the raised ring
     (around the plate) of an
3   astrolabe, in such fashion that the plate
     move in it and not become separated,
4   and the surface of the ring and the surface
     of the plate should form a single plane sur-
     face,
5   and in such fashion that from the circumference
     of the plate a narrow tongue
6   be extended. And in all the concavity of the
     ring let a hollow be cut out so that the
     tongue
7   fit into it. The flatness of this should be
     tested with ruler and plumbline,
8   as is known to the craftsman. Then, around
     the center of the disk,
9   draw five circles on the ring, and divide the
     first circle into twelve
10   parts, so that the fivefold circles are divi-
     ded by them,
11   and between the first and the second write
     the names of the (zodiacal) signs,
12   and consider the succession from the left,
     since it has itself been
13   assumed around the center of the disk. The
     second is divided into seventy-

ود و قسم کرده عدد درجات را بحکام ه در مابین ثانی

و ثالث رقم کنند و ثالث را ببیصد و شصت بعد د

درجات فلک البروج قسمت کنند و راج را بکسور بر

درجات بقدر امکان قسمت کنند و خطوط وصل

کنند میان راج و خاس بجهت کسور و خطوط هم

قسمی از اقسام حاوی تا دائن که اندرون هست

کنند و اگر خطوط اقسام خسات بلون دیگر کنند

بجهت تسهیل اعال هرترمو و پس روائر خامس که

بکسور منقسم است برماذاة هر قسمی ثمنی و قمن

کنند جا که مشتبه از حلقه بگذرد و زیابد قرص نیز

برماذاة او سوراخ شود ذهب انقلاب باید که بر

محیط یک ان بود تا بساز ند که کجر یک از ان

بس حصاری ی باریکه

انقلاب که فرو رود حلقه بر قرص ثبت شود

f.6v

1   two parts, the number of five(-degree intervals around the circle). Between the second

2   and the third place numerals. And the third (circle), divide it into three hundred and sixty, the number of

3   the degrees of the zodiac, and the fourth into fractions

4   of degrees to the amount possible, and join lines

5   between the fourth and the fifth for the fractions, and the lines of each

6   of the inner parts, extend (them) to the innermost circle.

7   If the lines of the five parts be made of different color(s)

8   for ease of operation it will be better. Then on the fifth circle, which

9   is divided into fractions, opposite each division, have a careful hole

10   so that the drill pass through the ring. And the tongue of the disk also

11   should be pierced opposite it, and all the holes must be along

12   the circumference of one circle. *Then* make *a very thin* peg so that whichever of those

13   holes it may pass through, the ring will be fixed on the disk,

---

* Marginal correction.

35

و باین حلقه قایم مقام فلک البروج بوده باشد

و هم درسم اوجات و مراکز و مناطق و نقطه

محاذاة و معدل المسیر برمحیط صفحه کیف ما اتفق

نقطه فرض کنند و اثر اوج شمس تامنه بس ازین

نقطه بمقدار مابین اوج شمس و اوج هر کوکب

برتوالی هر برج احسنذ کرده ۰ علامات بنهند

مابین اوج آفتاب و کواکب خمسه					
زحل	مشتری	مریخ	زهره	عطارد	
و ب م	ا ط ط	نو ه	ا بطا یو	سکطا نو ی	یح کع

بس میان هر علامت و مرکز بخطی مستقیم وصل کنند

چنانکه بعد از تمام آلت محاذات ممکن باشد و از مرکز

صفحه بسوی اوج هر کوکب غیر از شمس و برای شمس

سوی حضیض او و برای قمر بسوی مبدا اجزا آلة

f.7r

1   and this ring shall stand in the place of the
zodiac. CHAPTER

2   TWO. On the Drawing of the Apogees, and Centers,
and Deferents, and the

3   Opposite Point, and the Equant. On the circumference of the plate, however it may befall,

4   assume a point, and call it the sun's apogee.
Then, from this

5   point, to the amount between the sun's apogee
and the apogee of each planet, having taken
(it)

6   in (the direction of the) succession of the
signs, place marks.

7	(Angles) Between the Apogee(s) of the Sun and the Five Planets				
8	♄	♃	♂	♀	☿
9	5ˢ10;28°	2ˢ29;16°	1ˢ16;5°	11ˢ19;15°	4ˢ2;40°

10   Then, between each mark and the center, join
a straight line

11   such that, after the completion of the instrument, erasure be possible. And from the center
of

12   the plate, in the direction of the apogee of
each planet except for the sun, and for the
sun

13   in the direction of its perigee, and for the
moon in the direction of the beginning of divisions,

بمقدار مابین المرکزین بکبهند و بر آن علامتے نهند

اثر آن باقی ماند و ان علامت مرکز حامل آن

کوکب بود و بدین تفصیل که هندکو می شود و ا بین

ابعاد مراکز حوامل اکواکب علی ان نصف قطر الصفیحة سون						
ڡحہ	ا'هم	نری	ح	ڡ	حمہ	۔ح
و ط	س ط	ع	غ	د ل د ه	ط ل د	ا ۔ ٮد

بمقدار یت که نصف قطر صفیحه شفت درجه باشد

پس هر کی از بین علامات امرکز ساخته بر آی هر کیک

از قمر و حل و مشتری و مح و زهره بین هابین

بعد ازاین رسم کند و ابین و دوایر مناطق حوامل بکشد

انصاف قطار حوامل علی ان نصف قطر الصفیحة سون						
حمہ	اس	رحی	ک	ا'مں	ح	۔ح
م ط	ٮ لح	ٮ	ٮہ لح	ب	ٮ ر	غ خ

f.7v

1   to the amount between the two centers, lay (it)
off, and place a mark on it that

2   its trace shall remain. And that mark shall be
the deferent center of that

3   planet, according to the particulars recorded
herewith. And these

4	Distances of the deferent centers of the planets (in units) such that half the plate diameter is sixty.						
5	☉	☽	♄	♃	♂	♀	☿
6	2;6,9	10;19	2;58	2;32	4;33	1;2	4;52

7   (units) are of such an amount that half the plate
diameter shall be sixty degrees.

8   Then, each one of these marks having been made
a center, for each one

9   of the moon, and Saturn, and Jupiter, and Mars,
and Venus, with this distance (i.e., radius)
draw a circle,

10   and these circles are the deferent heavens of
these

11	Halves of the deferent diameters (in units) such that half the plate diameter is sixty.				
12	☽	♄	♃	♂	♀
13	49;41	52;2	55;28	45;27	58;58

کواکب است و احتیاج بر سم منطقه خارج شمس
نیست چه خط صفحه منطقه شمس فرض کند مستفار
و مرکز مرسوم را مرکز مستعار شناسا نند و آن
مرکز که برای عطارد رسم کرده اند مرکز مدیر عطارد
باشد پس از مرکز مدیر خطی اخراج کند مقاطع
خط و حجی بر قایمه و از ین مرکز بهره و جانب یمین
و یسار خط و حجی بمقدار ه ح دو علامت بنهد
و هریک ازین علامتین را مرکز ساخت بعد
تا کح و قوس رسم کند تا شکلی اهلیلجی حاصل
شود که نصف قطر طول و ناح نصف قطر
اقصرش مو ه بود و این مقدار مرکز دنگ و یر عطارد
بود و ها اورا منطقه عطارد نامیم و باید که هریک
از منطقه کو کبی را بدونی دیگر کشند تا در وقت عمل

f.8r

1   planets, and there is no need for drawing the
eccentric heaven of the sun,

2   since the circumference of the plate is ficti-
ciously assumed (to be) the heaven of the sun,

3   and the marked center (of the plate) is called
the ficticious (deferent) center of the sun.
And that

4   center which has been drawn for Mercury, let
it be Mercury's turning center.

5   Then, from the turning center extend a line
intersecting

6   the line of apsides perpendicularly, and from
this center to both sides, right

7   and left of the line of apsides, to an amount
5;8, make two marks,

8   and, having made each one of these two marks
a center, with distance (i.e., radius)

9   51;23 draw two arcs so that an elliptical-
shaped figure results,

10   of which half its larger diameter is 51;8 and
half its

11   shorter diameter is 46;15, and this is the
orbit of Mercury's epicycle center,

12   and we call (it) Mercury's deferent. And each

13   of the planet's deferents must be drawn in a
different color, so that at the time of operation

شتبه کرد و بس از مرکز منطقه هر یک از علویه و زهره
بجای الحجار در قسم از مرکز از مرکز صفیحه جانب نخا
مبدا احسبا یعنی بطرف میزان بقدر بعد مرکز آن
کوکب از مرکز صفیحه نکنه ند و بران علامتی بنهد
کراخران با قی ماند و در عطار و برمنتصف مابین مرکز
صفیحه و مرکز دیر علامت کند و این علامت را
در غیر قمر مرکز معدل النهسر و در قمر نقطه محاذاه خواند
باب سیوم در رسم قط استوا و نقطه عرضی
و خطوط طاعرض و مبادی نطاقات و صورت صفیحه
و ناطق و علامات و دواير و خطوط بر صفیحه قطری
رسم کنیم که مبدا اجسا و محیط کذ رده و دايره نقطه
استوا نا میم بس بر قطر استوا بقرب اول
میزان هشت علامت کنیم چنانکه اثر علا ات

f.8v

1   no mistake be made. Then, from the deferent
center of each of the superior (planets), and
Venus

2   on the side of its apogee, and for the moon from
the center of the plate on the opposite side
from

3   the beginning of divisions, i.e. on the side
of Libra, to the amount of the distance of the
center of that

4   planet from the plate center, lay (this distance) off, and put a mark on it

5   that its trace remain. And for Mercury, on the
half(-distance) between the center of

6   the plate and the turning center make a mark,
and these marks,

7   except for the moon, are the equant centers,
and for the moon it is called the opposite
point.

8   CHAPTER THREE. On drawing the Equating Diameter, and the Latitude Point(s),

9   and the Latitude Lines, and the Beginning of
the Sectors, and the Picture of the Plate,

10 and the Deferents, and Marks, and Circles,
and Lines. On the plate draw a diameter

11 which passes through the beginning of the divisions and the circumference, and we call it
the equating diameter.

12 Then, on the equating diameter, near the first
(point) of

13 Libra, we make eight marks such that the traces
of the marks

ابعاد خط عرض اکوکب الخمسة علی ان نصف قطر الصفحه سبعون				
زحل	مشتری	مریخ	زهره	عطارد
خ ج د ۱۵	ن و ح و	خ ط ط ۲ ۱۵	مذلح لح ح ۲	صو ؟

وبدین علامات را نقطهٔ عرض خوانیم واگر برای
هریک از علامتی نیز نقطه علامات واحدها کتفا کند
در منتصف بعد بین در مقصود خلل نذ بد پس
برکز صفحه نصف ابع رسم کنیم بر واحد بین قطر
استوا اگر بر جانب بروج جنوبی بود بهتر که
نصف قطر او مساوی حب نذ درج بکشد درمقط
اجر آن مجره بعد از ان سطان را اب هر دو

T R A N S L A T I O N

f.9r

1 remain, two for Saturn, and two for Jupiter,

2 and two for Mars, one for Venus, and one for Mercury.

3 And the distance of each mark from the center of the plate is detailed herewith:

4	Distances of the latitude line(s) of the five planets (in units) such that half the plate diameter is sixty							
5	♄		♃		♂		♀	☿
6	53;55	50;1	57;46	53;9	50;0	40;54	58;58	46;0

7 and these marks we call latitude point(s), and if, for

8 each of the superior (planets) also one is content with

9 halving (the sum of) the two distances the objective will not be marred. Then,

10 with the center of the plate (as center) draw a semicircle on one side of the equating diameter,

11 if on the side of the southern signs (it is) better, such that

12 half its diameter be equal to the sine of nine degrees of the concavity

13 of the ring. After that, place a ruler upon pairs of

In the case of the inferio (planets) the mark(s) have been made single, because they do not have two extremes, for the epicycl center of Venus is always north and of Mercury alwa south.

This is the distance of the planet's epicycle center from the center of the universe in northern and southern directions.

جزء مساوی لبعد از قطر استوا، بهند
خطی می کشند در داخل نصف دایره خطوط استوا ....
که موازی خط استوا بود و منتهی شود و لا محال ابعاد
بین این خطوط به نسبه جیوب قسی خواهد بود از یکی
تا نود بر محیط نصف این اعداد خطوط را استوا یله
از جانبین می کشند و همچنین خطوط دقایق ربع دایره
نه امکن رسم کنند و این خطوط را خطوط عرض نامیم
و همچنین بر مرکز صفحه نصف دایره رسم کنند
که مماس خط یح درجه از خطوط عرض یح را این ....
و آیره عرض قمر خوانیم و اگر صفحه بزرگ باشد
رسم خطوط بدقیقه دقیقه یا دقیقتین دقیقتین کرد ... باشند
دو دایره دیگر رسم کنند یکی مماس خط ٥٠ دقیقه
و آزا دائن عرض اول ترسیم خواهیم و دیگری

f.9v

1   points equidistant from the equating diameter,

2   and draw line(s) inside the semicircle *(up)
    to the semicircle* in parallel lines

3   which are parallel to the equating diameter
    (and such that the inside) be filled. And
    undoubtedly the distances

4   between the lines will be in the ratio of the
    sines of the parts from one

5   to nine. Then on the circumference of the
    semi-circle, write the numbers of the succes-
    sive lines

6   from both sides, and in the same fashion draw
    also the lines of the minutes

7   (to the amount) possible, and these lines we
    call latitude lines.

8   And in the same manner, draw a semicircle about
    the center of the plate

9   which shall be tangent to the five-degree line
    of the latitude lines.

10 We call this the moon's latitude circle. And
    if the plate is large

11 the lines may have been drawn minute-by-minute
    or (every) two minutes.

12 Draw two more circles, one tangent to the ten-
    minute line,

13 and we call it the circle of Venus' first
    latitude; and the other

---

*Marginal addition.

47

کمامن خط چهل چهل و پنج دقیقه وان را
دائره عرض عطارد نامیم پس بر منطقه
هر درجه که نقطه چهار علامت کنند دو
براوج و حضیض و دو بر و مبدأ نطاق
نامیم و رابع محبت البعد بجهت شهرتش
وسقه اربعه این دو مبدأ از جدول
ربع آخذ کنند وما جدولی اور دیم که مبدأ
نطاقات محبت البعد والحمر که تفسیر
کدام که خواهند از ان جدول
برد ارند وآن جدول
اینت

f.10r

1   tangent to the forty-five minute line, and we call it[*]

2   the latitude circle of Mercury. Then, on the deferent of

3   each planet make four marks, two

4   on the apogee and perigee, and two on the two beginnings of the second and fourth sectors

5   reckoned according to distance, as is known (for each particular planet).

6   And the amount of the distance of these two beginnings

7   is taken from a table of the (Khāqānī) Zīj, and we have brought (out) a table such that

8   the sectors, computed according to distance and motion, any

9   which is wanted, take it from that table,

10   and that table

11   is this:

---

[*]The word اخرى is repeated in the text.

| جدول مبادى نطاقات | | | | | | | | |

نطاق الاوجى بالمركز — نطاق التدويرى بالحاصة المعدلة

الكواكب		نطاق الاوجى بالمركز				نطاق التدويرى بالحاصة المعدلة		
	اول	ثانى	ثالث	رابع	اول	ثانى	ثالث	رابع
	؟؟؟				؟؟؟			

f.10v

TABLE OF THE BEGINNINGS OF THE SECTORS

The Planets		Apogee Sectors, (according) to the Center				Epicycle Sectors (according) to the Adjusted Anomaly			
		First	Second	Third	Fourth	First	Second	Third	Fourth
According to Distance	☉	0³0;0°	3³1;0°	6³0;0°	8³29;0°	0⁶0;0°			
	☾		2 24;2		9 5;58		3⁵2;31°		[8]⁵27;2[9]°
	♄		3 4;54		8 25;6		3 2;56		8 27;4
	♃		3 3;26		8 26;34		3 [5];3[8]		8 2[4];22
	♂		3 8;30		8 21;30		4 17;48		7 12;12
	♀		3 1;1		8 23;59		3 20;43		8 9;[17]
	☿		2 7;44		9 22;16		3 9;[2]3		[8] 20;37
According to Motion	☉		3 2;1		8 28;[59]				
	☾		3 24;0		8 6;0		3 5;3		8 24;57
	♄		3 3;16		8 26;44		3 5;53		8 27;7
	♃		3 2;37		8 27;23		3 10;34		8 19;26
	♂		3 5;42		8 24;18		4 7;38		7 22;22
	♀		3 1;[0]		8 29;0		4 15;0		7 15;0
	☿		3 3;2		8 26;58		3 19;1		8 10;59

صورت حجره وصفحه ومساطق وعلامات ودواير
وخطوط اینست

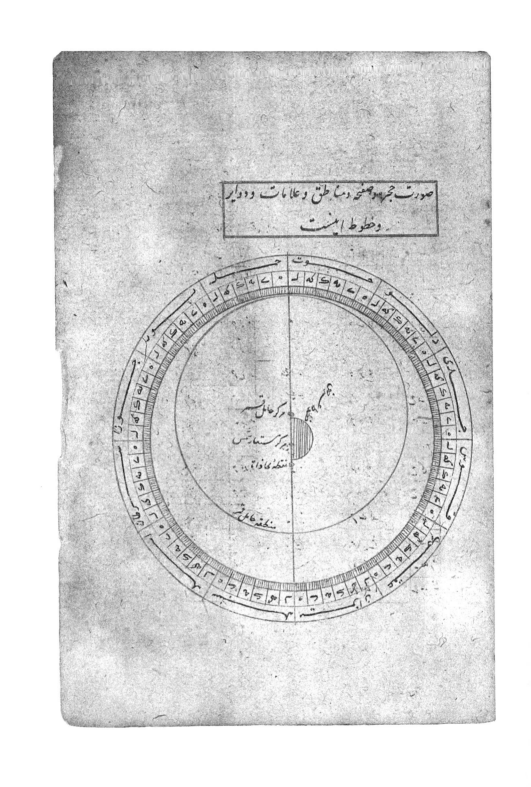

f.11r

The Picture of the Ring and Plate and
Deferents and Marks and Lines is this:

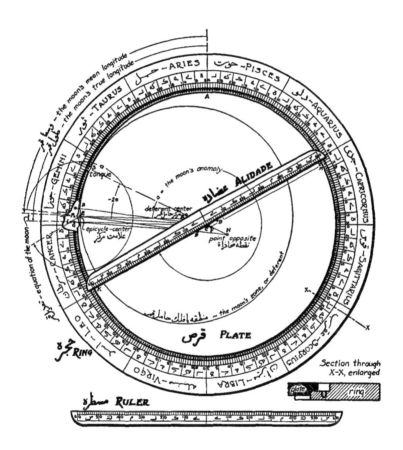

(Figure 1)

باب چهارم در صنعت عصا و
مسطره دوسطره باز ندانخاس یا صفحه
یا خشب یکی مثل عصاده اصطلاب که اورا
حرف واحد باشد و باید که طول آنزیاده باشد
ازقطر صفحه واقصر ازقطر محدب حجره
وحرف اورا از مرکز نامقدار نصف قطر
صفحه از جانبین بشت کنند مثل عصاده
اصطلاب محب واعداد اجزاآن و اورا بر وسو
جانب مرکز طرد او عکسا بنویسند وباید که
موضع دور قطب درنصف دائره باشد
ودر تصویر آن زائده بقدر امکان کو کشند تامرکز
بیداری درشیب او مختفی نماند ودیگری پیز
بهان مقدار واجز۰ لکن بجای زائده بطر

f.11v

1   CHAPTER FOUR. On the Construction of the Ali-
      dade and

2   Ruler. Make two rulers, of copper, or brass,

3   or wood, one like the alidade of an astrolabe,
      such that (it) shall be

4   one (straight-)edge, and its length must be
      in excess

5   of the diameter of the plate and shorter than
      the convex diameter (i.e., outer diameter) of
      the ring.

6   And its edge, from the center to the amount of
      half the diameter of the

7   plate, from both sides divide it into sixty
      parts, like the alidade of a

8   sine astrolabe, and the numbers of its divi-
      sions are written proceeding in opposite

9   directions from the center, and the

10   place where the pivot passes (through it) must
      be (on a projecting) semicircle,

11   and in making that (semicircular) projection
      as small as possible, care should be exercised
      that the center

12   not be hidden in its slope. And the other
      (ruler) is also

13   (made) the same size, and with the same divi-
      sions, but in place of the projection on the

اول در وجهی مستدیر کند که زائده اول
عند الحاجت در و در آید و هره و مسطره چون آن
مسطره واحد کرد و اعداد مسطره ثانیه
را از یک طرف تا طرف دیگر بنویسد
طرد او عکسا و با مسطره اول را عضاده
و سطاه ثانیه را مسطره خائیم پس
بر احد جانبین عضاده کشش علامت کند
بجهت کواکب سته غیر شمس و ازار نام
اختلاف نائیم و ابعاد علامات از مرکز صفحه
برای هر کوکبه بدین فصل است

قطر الصفحه سوا	علی ان نصف	ایام الاختلاف	ابعاد
قمر	زهره	مریخ	مشتری	شمس	عطارد	زحل
و	ل ح	ل	ل ا	ل ا	مسکه	ل ح	ط

f.12r

1   first ruler make a (semi)circular depression
in it such that the projection on the first,

2   when needed, may enter it, and both rulers
become

3   as one ruler. And the numbers of the second
ruler

4   are to be written from one side to the other,

5   outward and in opposite directions. And we
call the first ruler the alidade,

6   and the second ruler we call the ruler. Then,

7   on one of the two sides of the alidade make
six marks

8   for the six planets other than the sun, and
we call these the

9   difference marks, and the distances of the
marks from the center of the plate

10   for each planet are detailed herewith:

Distances of the difference marks in 60ths of half the plate diameter:					
☾	♄	♃	♂	♀	☿
5;17	5;38	10;38	30;32	42;25	18;13

11
12
13

This distance is to the amount of half the epicycle diameter of these planets.

57

واگر در صفحه بر مرکز صفحه ششش دائره رسم کنند
بدین بعد برای اختلاف رو ابعد ان دوایر
اگر اختلاف نامند واز اخذ راسین مسطره
تا بد یعنی از محدد راس و بعد از مبداء جبذا
به بهشتت و سه جبزا از جوا جبذا مسطره
بر برکار کرفته جیث بلغ بر حرف اوعلامتی کنند
واگر علامت خسوف نامند و بعد کسی و
جبزا علامتی دیگر کنند واگر علامت کسوف
خوانند و به بعد پست و به علامتی دیگر
کند بجهت مکت خسوف و بر وجه مسطره
زنهایت جزا تاسع و عشرین تانهایت جزا ثالث
وستین در طول مد وازده قسمت کنند بجهت معرفت
اصابع منخسفه و از مبداء جوا مسطره از طرف

f.12v

1 And if on the plate, around the plate center,
   six circles are drawn

2 with these distances (as radii) for difference(s),
   it is permissible, and those circles

3 are called difference circles. And from one
   of the two heads of the second ruler,

4 i.e., from beyond its head, not from the beginning of divisions,

5 to a distance of sixty-three divisions of the
   divisions of the ruler,

6 laid off with a compass, wherever it reaches,
   on its edge make a mark,

7 and that is called the lunar eclipse mark. And
   at a distance of thirty-three

8 divisions make another mark, and it is called
   the solar eclipse mark.

9 And at a distance of twenty-nine make another
   mark

10 for the duration (or first totality) of the
   lunar eclipse, and on the face of the ruler,

11 between the limits of the twenty-ninth division to the sixty-third division,

12 divide in length (the segment thus determined)
   in twelve (parts) for the determination of

13 lunar eclipse digits. And from the beginning
   of the divisions of the ruler, on the other
   side,

دیگر تا نهایت جزء ثالث و ثلثین بد وازده
قسمت کنند بجهت معرفت اصابع تنکسف و
اعداد اصابع در هر ره و از جهت مرکز بنویسند
و مسطرتین را السلسله و قبقه مرتبط کرده
که طول سلسله قریب باشد بطول نصف قطر
و نصب لبنتین در عرض داده در اخذ ارتفاع بکار آید
و اگر بجای مسطره ریسمانی باریک استعمال
کنند مطلوب حاصل کرده و باب چهارم
در رسم جداول اوساط و غیرها بر ظهر صفیحه
جدولی رسم کنند که در عرض بیازد و قسمت
کرده بکشند یکی از جهت صف اعداد و پنج بجهت
اوساط طبرین و علویه و پنج باقی بجهت اوج
شمس و خاصه قمر و وسط جوزهر و

f.13r

1   to the end of the thirty-third division divide (it) into

2   twelve parts for the knowledge of the solar eclipse digits, and write

3   the numbers of the digits of both, (beginning) from the center.

4   And let the two rulers be joined with a thin chain

5   such that the length of the chain be nearly the length of half the diameter.

6   And the placing of two sights on the alidade may be used for taking the altitude.

7   And if, in place of the ruler, a thin string is used,

8   the desired result may be attained. CHAPTER FIVE

9   On Drawing the Tables of Mean (Motions), etc. On the back of the plate

10   draw a table whose width is divided into eleven parts,

11   one for the column of numbers, and five for

12   the mean (motions) of the sun and moon and superior planets, and the remaining five for the

13   solar apogee, and the lunar anomaly, and the mean of the lunar nodes, and

و خاصه مرکب یسطیین و در طول به پنجاه و شت

قسم مستقیم سه جهت مسطور القاب و ده برای

حرکت اوساط در ده سال متوالی از سنین ما قصه

یزد جردی و نوزده بحسب سنین عشرات و آحاد

و الف و سیزده برای شهور اثنی عشر

خمسه و دوازده برای حاد و عشرات الایام و یکی برای

ساعت و جدولی دیگر رسم کنند در عرض مستقیم

چهار قسم یک صف بجهت اسماء کواکب

خمسه متحیره و یکی برای مبدأ حصه در مقالات

رجعت و دیگری منها ی حصه و د مقالات

رجعت و یکی دیگر برای تفاضل منها

و ما این جدول را در باب نهم در معرفت

رجعت و استقامت او ردیم و اگر این

f.13v

1   the compound anomaly of the two inferior planets;
and in length (it) is divided into fifty-eight

2   parts: three for the column headings, and ten
for

3   the motion of the mean (planets) in ten successive years of incomplete Yazdigerd years,

4   and nineteen on account of the tens and hundreds
and a

5   thousand, and thirteen for the twelve months
and

6   five (intercalary days), and twelve for single
(days) and tens of days, and one for the

7   hour. And lay out another table (to be) divided

8   into four parts, one row for the names of the

9   five planets, and one for the beginning of the
limits of the

10   retrograde stations, and another (for the)
ends of the boundaries of the

11   retrograde stations, and one more for the
differences between the two of them.

12   And we have brought forth this table in the
ninth chapter, on the determination of

13   retrogradations and stations, and if this

جدول بارسم نکنند و برای هر جزءِ حجره برابر یک

بها دی حد و د استقامت و نهایات ان علامت

نهند بهتر و یک جدول و یکمر وضع کنند

بجهت اختلاف ساعات و انها این اجتماع

حقیقتی و مرئی بود و اختلاف منظر قمر

در عرض عند الاجتماع المرئی بعرض

وسط اقالیم و ما این جدولها را از زیج خاقانی

عمل کرده نوشتیم و اکر این جدولها را در آنکه

وضع کنند و عند الاحتیاج از این رساله

بردارند جایز است و جدول اختلاف ساعات

چون در معرفت کسوف بکار می آید در باب

معرفت کسوف نوشتیم و صورت جدول

اوساط اینست و الله اعلم

f.14r

1   table is not drawn, and on the divisions of
the ring marks are put for

2   the beginning of the boundaries of the stations and their ends,

3   (it will be) better. And place another table

4   for the difference of the hours, and that will
be between the

5   true and apparent conjunction, and the lunar
parallax

6   in latitude according to the apparent conjunction at the

7   mean latitude(s) of the climates. And we,
having made (?) these tables in the Khāqānī
Zīj,

8   wrote them (from it). But if these tables are
not

9   put on the instrument, and, when necessary,
are taken from this treatise,

10   it is permissible. And the table of difference
of the hours,

11   since it was being used in the science of
solar eclipses, we wrote it in the chapter
(on)

12   the science of solar eclipses. And the picture
of the table of

13   mean motions is this, but God knows better.

عطارد		زهرة	مريخ	مشترى	زحل		القمر			الشمس	
خاصة	وكبه	خاصة	وسط	وسط	وسط	وسط الجوزهر	الحاصة	الوسط	الاوج	الوسط	

f.14v   (A Table of Mean Positions
                 and Mean Motions)

Years, months, days, & hours	☉ Mean	☉ Apogee	☾ Mean	☾ Anomaly	☊	♄ Mean	♃ Mean	♂ Mean	♀ Compound Anomaly	☿ Compound Anomaly
851	8ˢ 4;22°	3ˢ 1;59°	6ˢ 1;49°	3ˢ 28;46	9ˢ 24;4°	6ˢ 12;7°	4ˢ 24;9°	6ˢ 4;8°	3ˢ 29;57°	7ˢ 8;4°
852	4;7	3 2;0	10 11;12	6 27;29	10 13;24	6 24;[2]1	5 24;29	0 15;25	11 14;44	9 1;48
853	3;53	1	2 20;35	9 26;12	11 2;43	7 6;[3]5	6 24;50	6 26;4[2]	6 29;32	10 25;32
854	3;39	2	6 29;58	0 24;55	11 22;3	7 18;[4]8	7 25;10	1 8;0	2 14;[0]9	0 19;16
855	3;24	3	11 9;20	3 23;38	0 11;23	8 [1;2]	8 [25;31]	7 19;17	9 29;6	2 13;0
856	3;10	3	3 18;44	6 22;22	1 0;43	8 1[3;16]	9 25;51	2 0;34	5 13;54	4 6;44
857	2;56	4	7 28;7	9 21;5	1 20;2	8 2[5];[19]	10 26;12	8 11;51	0 28;41	6 0;28
858	2;41	5	0 7;30	0 19;49	2 9;22	9 [7;43]	11 26;33	2 23;8	8 13;29	7 24;12
859	2;27	6	4 16;53	3 18;31	2 28;42	9 [9;5]7	0 26;53	9 4;25	3 28;16	9 17;55
860	8 2;13	3 2;8	8 26;16	6 17;14	3 18;1	10 [2;1]0	1 27;14	3 15;43	11 13;4	11 11;39
10	11 27;37	0 0;9	7 3;51	5 17;11	[6] 13;17	4 2;17	10 3;26	3 22;52	2 27;54	5 27;19
20	25;13	17	2 7;4[2]	11 4;23	0 26;35	8 4;33	8 6;52	7 15;44	5 25;49	11 24;38
30	22;50	26	9 11;33	4 21;34	7 9;52	[0] 6;50	6 10;17	11 8;35	8 23;43	5 21;27
40	20;26	34	4 15;24	10 8;46	1 23;9	4 9;6	4 13;43	3 1;27	11 21;38	11 19;16
50	18;3	43	11 19;15	3 25;57	8 6;26	8 11;23	2 17;9	6 24;19	2 19;32	5 16;56
60	15;40	0 0;51	6 23;4	9 13;9	2 19;43	0 13;39	0 20;35	10 17;11	5 17;26	11 13;55
70	13;17	0 1;0	1 26;58	3 0;20	9 3;0	4 15;56	10 24;0	2 10;3	8 15;21	5 11;14
80	10;53	9	9 0;49	8 17;32	3 16;17	8 18;12	8 27;26	[6] 2;54	11 13;16	11 8;33
90	8;30	17	4 4;40	2 4;43	9 29;35	0 20;29	7 0;52	9 25;46	2 11;10	5 5;52
100	11 6;6	0 1;26	11 8;31	7 21;55	4 12;52	4 22;46	5 4;18	1 18;38	5 9;4	11 3;11
200	10 12;13	0 2;51	10 17;1	3 13;49	8 25;43	9 15;32	10 8;36	3 7;16	10 18;9	10 6;22
300	9 18;20	4;17	9 25;32	11 5;44	1 8;35	2 8;17	3 12;53	4 25;54	3 27;13	9 9;33
400	8 24;27	5;43	9 4;3	6 27;38	5 21;27	7 1;3	8 17;11	6 14;32	9 6;18	8 12;4[5]
500	8 0;34	7;9	8 12;34	2 19;33	10 4;19	11 23;50	1 21;29	8 3;10	2 15;22	7 15;56
600	7 6;41	8;34	7 21;5	10 11;28	2 17;10	4 16;35	6 25;47	9 21;48	7 24;27	6 19;7
700	6 12;47	0 10;0	6 29;35	6 3;22	7 0;2	9 9;20	0 0;4	11 10;27	1 3;31	5 22;18
800	[5] 18;54	11;26	6 8;6	1 25;17	11 12;54	2 2;6	5 4;22	0 29;5	6 12;36	4 25;20
900	4 25;1	12;51	5 16;37	9 17;12	3 25;46	6 24;52	10 8;40	2 17;43	11 2[1];40	3 28;40
1000	4 1;8	0 14;17	4 25;7	5 9;6	8 8;37	11 17;38	3 12;57	4 6;21	5 0;45	3 1;52
Farvardin	0 0;0									0 0;0
Ordibehesht	0 29;34	0 0;0	1 5;18	1 1;57	0 1;35	0 1;0	0 2;30	0 15;43	1 18;4	4 2;46
Khordad	1 29;8		2 10;35	2 3;54	3;11	2;1	4;59	1 1;27	3 6;8	8 5;33
Tir	2 28;42		3 15;53	3 5;51	4;46	3;1	7;29	1 17;10	4 24;12	0 8;19
Mordad	3 28;17		4 21;10	4 4;48	6;21	4;1	9;59	2 2;53	6 12;16	4 11;5
Shahrivar	4 27;51		5 26;[27]	5 9;45	7;57	5;2	12;28	2 18;37	8 0;19	8 13;52
Mehr	5 27;25		7 1;45	6 11;42	9;32	6;2	14;58	3 4;20	9 18;23	0 16;38
Aban	6 26;59	0 0;1	8 7;3	7 13;39	11;30	7;2	17;27	3 20;3	11 6;27	4 19;24
Adhar	7 26;33		9 12;20	8 15;36	12;42	8;2	19;57	4 5;47	0 24;31	8 22;11
Dey Mah	8 26;7	1	10 17;38	9 17;33	14;18	9;3	22;26	4 21;30	2 12;35	0 24;57
Bahman	9 25;42	1	11 22;55	10 19;30	15;53	10;3	24;56	5 7;13	4 0;39	4 27;43
Isfandarmadh	10 25;16	1	0 28;13	11 21;27	17;18	11;3	27;26	5 22;57	5 18;43	9 0;30
Five days	11 24;50	0 0;1	2 3;30	0 23;24	0 19;4	0 12;4	0 29;56	5 8;40	7 6;47	1 3;16
1	0;0	0 0;0								0 0;0
2	0 0;59		13;11	0 13;4	0 0;3	0 0;2	0 0;5	0 0;31	0 1;36	0 4;6
3	0 1;58		0 26;21	0 26;8	6	4	10	1;3	3;12	8;11
4	0 2;57		1 9;32	1 9;12	10	6	15	1;34	4;48	12;17
5	0 3;57		1 22;42	1 22;16	13	8	20	2;6	6;24	16;22
6	0 4;56		2 5;53	2 5;19	16	10	25	2;38	8;1	20;28
7	0 5;55		2 19;3	2 18;23	19	12	30	3;9	9;37	24;33
8	0 6;54		3 2;14	3 1;27	22	14	35	3;40	11;13	28;39
9	0 7;53		3 15;25	3 14;31	25	16	40	4;12	12;49	1 2;44
10	0 8;52		3 28;35	3 27;35	29	18	45	4;43	14;25	1 6;50
10	0 9;51		4 11;46	4 10;39	0 0;32	20	0 0;50	5;14	0 16;1	1 10;55
20	0 19;45		8 23;32	8 21;58	0 1;4	0 0;40	0 1;40	0 10;8	1 2;2	2 21;[5]1
The hour	0 0;2	0 0;0	0 0;33	0 0;33	0 0;[0]	0 0;[0]	0 0;0	0 0;1	0 0;4	0 0;10

کاغ‍ذ در صفت لوح انتقال است و

صورت آن که هذا لوحی از خشب صلب شد

یا صفر که اطولش یک ذراع و عرض آن شتر

لشان دو ذراع و هر کستوا سطح لوح مقدر با ممکن سنبیه

کند و پر و حسب لوح منتفع قائم الزاویه بر رک

کند چنانکه یک از وزنه قائمه

شکلت موازای ذو صلع لوح بود کل نظیه ببندی

اصلاح لین صلع اطول زاویه قائمه راکه قاعده

شکلت است بیست و چهار قسم کند برای

ساعات و هر قسمی را استبصت با آن قدر

که توانند بحسب صفر که لوح قسمت کنند

و صلع نشانترو و قسمت کم و هر قسمی را ابسطت

یا یا نجه توا نند قسمت کنند و اکر قائم مراتب

القمرا

f.15r

1 CONCLUSION, On the Construction of the Plate
of Conjunctions. And

2 (so) get its picture; a plate of hard wood, or
yellow copper,

3 or brass, that its length be one cubit and its
width more than

4 two thirds of a cubit. And, to the extent
possible, attempt to make the surface of the
plate flat.

5 And on the face of the plate draw a right
triangle

6 such that each one of the two legs of the

7 triangle be parallel to two sides of the plate
respectively, at a proper distance.

8 Then the longer leg of the triangle, which is
the base

9 of the triangle, divide into twenty-four parts
for

10 the hours, and each part is divided into sixty
(parts) or to the amount

11 which it can be (divided), taking into account
the smallness or largeness of the plate.

12 And the *shorter* leg, having been divided
into sixteen parts, each part is divided into
sixty,

13 or to the extent to which it can be divided.
And if the divisions of the (planetary) paths

The method of verifying the right angle is this, with a compass from one of the two sides of the right angle, lay off three equal parts, and from the other side, with just the same opening of the compass, lay off four parts, and from the end of the third part to the end of the fourth part join a line so that it is the chord of the angle. If that line, with the same opening of the compass, is five parts (the angle is verified), and otherwise it is not a right (angle). Its proof is that the squares on the two sides of the triangle are equal to the square of the chord of that triangle, as angle of the right triangle, as is explained in (the Theorem of) the Bride (the Pythagorean Theorem).

I.e., in the width of the plate have that amount outside the triangle that the number(s) of travel (can) be written, and in the length have that amount that the parts and amount that the rulers and numbers of hours may find place.

---

*Marginal correction.

اقسام ساعات برابر نباشد در بزرگ
وجود و بعض بود و استحسانا لاوجو پس
ازین هر قسمی از اقسام صلین خطوط اخراج کنند
موازی ضلع آغز متصل شود بخطی
که از قسام آن ضلع دیگر موازی این ضلع
اخراج کرد و پسکند و تمیز میان خطوط ظاهر
از مقاسم ساعات و اجزاء میسرات و خطوط
ظاهر جسم از دقایق هر یک با لوان مختلفه
کنند و همچنین خمسات اجزا و دقایق را
نیز بلونی مغایر آنها تمیز کنند که عندالعمل
آسان کرده و خارج ضلع اطول ثلث
حفری کنند سر تاسر لوح که عرض در یک مقدر
اصبعی نبشد و عمقش بمقدار بی صالح بود و

f.15v

1   be equal to the divisions of the hours, in
    largeness

2   and smallness (of divisions), it will be
    better; adornment is unnecessary. Then,

3   from each of the divisions of the two sides,
    extend lines

4   parallel to the other side until they join the
    line(s)

5   which have been extended from the divisions
    of that other side parallel to this (first)
    side.

6   And (for) distinguishing between the lines
    extended

7   from the divisions of hours and parts of
    (planetary) paths, and the lines

8   extending from the minutes, make each in
    different colors.

9   And in the same fashion, the fifths of parts
    and minutes

10 also distinguish them by a color differing
    from those (others), that operation (with it)

11 be facilitated. And outside the longer side
    of the triangle

12 make a trough from one side of the plate to
    the other, that its width be to the amount of
    a

13 digit and its depth be to a proper amount,
    and

واندرون حفره را از جانب قاعده مثلث بزرگتر کنند
و در پهلوی این حفره حفره‌ی دیگر کنند که عرضش و عمقش
با حفره اول برابر بود و طولش از حد آن زاویه
قایمه تا ثلثه ارباع قاعده مثلث بود و درون
این حفره از جانب حاشیه لوح بزرگتر سند
واگر این حفره را با حفره اول متصل کنند شاید پس
سه مسطره بسازند از خشب یا صفر
که غلظ هر یک بمقدار وسعت حفره بود و طول
یکی ازین مساطر بمقدار ثلث قاعده مثلث بود
واین را مسطره غدات نامیم و طول هر یک
ازان دو مسطره دیگر بمقدار ثلثان قاعده مثلث
باشد یکی را مسطره یوم و دیگری
را مسطره لیل خوانیم پس مسطره لیل را

f.16r

1   make the inside of the trough bigger than (it
     is) beside the base of the triangle.

2   And beside that trough, make another trough
     such that its width and depth

3   be equal to (those of) the first trough, and
     its length

4   (as compared to) the limit of that right *angle*
     of the triangle should be three fourths the
     base of the triangle, and the inside

5   of this trough is made bigger than the (width)
     alongside the margin of the plate.

6   And if this trough is joined with the first
     trough, perhaps (it will be satisfactory).
     Then

7   make three rulers, of wood or brass,

8   that the thickness of each one to be the amount
     of the width of the trough, and the length of

9   one of these rulers be to the amount of a third
     of the base of the triangle.

10   And we call this the next-day ruler, and let
     the length of each

11   of these other two be to the amount of two
     thirds of the base of the triangle.

12   One of these we call the day ruler, and the
     other

13   the night ruler. Then put the night ruler

---

*Marginal addition.

در حفره‌ای وسطه غداة و یوم

را در حفره اول کنند و سنی که سطه

غداة از جانب زاویه فایده بود و این

سطه درین حفرها حرکت کند و ازوجه

لوح مرتفع نکردند و سطوح خاص سطه

با سطح لوح بمشابه یک سطح مستوی کرد د

و به سطح را با جزا فاع مثلث د

د قابض قسمت کنند پس سطه غداة بهشت

سم و آن دو ی دیگر شانزده قسمی و

داد قام اعداد هر سطح از جانب زاویه

عاده بسوی زاویه قایمه نویسند

و درمابین حفره و محط لوح اعداد ساعات

وخسات دقایق قاعی سطح مثلث

f.16v

1   in the second trough, and the next-day and
day rulers

2   put in the first trough in such fashion that
the

3   next-day ruler be from the side of the right
angle. And these

4   rulers must be able to move (i.e. slide) in
these troughs, but should not be (capable of
being) raised above the surface of the

5   plate, but the visible surfaces of the rulers

6   should be like a single plane surface.

7   And divide the face(s) of the ruler(s) into
the divisions of the base of the triangle and

8   its minutes. So the next-day ruler (will be
divided) into eight

9   parts, and those other two (into) sixteen, and

10   write the signs of the numbers of each ruler
from the side of the

11   acute angle along the base of the triangle.

12   And between the trough and the perimeter of
the plate, write the numbers of the hours

13   and of the fives of minutes of the base of
the triangle

از جانب زاویه حاده بنویسند تا زاویه
قایمه و اعداد اجسزا، سیرات و
خسات و قاین آن را از زاویه قایمه
سرها تا آخر اجزا بنویسند در محدد
زاویه حاده که اثبات اعداد سا عانکه
از یک کرده شده است ثقبی و قیق کند
و خطی از ان ثقب که زا ان
بمقدار و ترز او زاویه قایمه یا برود جو لوح حظره
حرف سازند بمقدار و بز مدکور و بحو
برثقب مدکور محکم کنند که آن
سطره بر آن محدد حرکت کند و ما
این مسطره را مسطره مدین
خاتم صدوق لوح است

f.17r

1   from the side of the acute angle to the

2   right angle. And write the numbers of the
divisions of (planetary) paths, and

3   the fives of their minutes from the right angle

4   from beginning to end. And at the extremity
of

5   the acute angle, which is from (where) the
beginning of the numbers of the hours,

6   has been made, drill a minute hole.

7   And make a thread to pass through that hole,
(in length)

8   to the amount of the hypotenuse of the right
triangle. Or, on the face of the plate emplace
an

9   edged ruler, to the amount of the (above-)
mentioned hypotenuse, and fasten it to an axis

10   on the (above-)mentioned drill(-hole) such
that that

11   ruler move about that axis. And we

12   call this ruler the turning ruler.

13   And the picture of the plate is this:

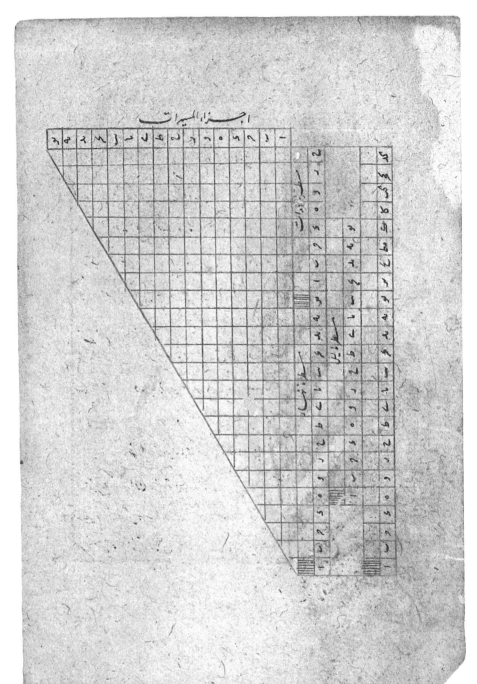

f.17v

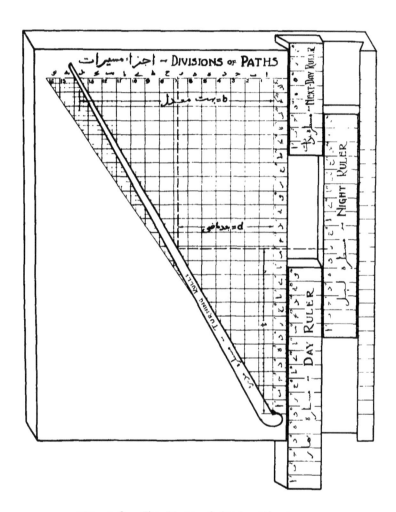

Figure 2. The Plate of Conjunctions

وصف له ... و جم ... و عمل نطق الناطق

مستقل بر باز دید . باب وخاتمه ... باب ... ول

در ترتیب اوج الخ شمس ام بر صفحه کرد . بود ... د

بر محاذی اوج شمس نهند اجزا ... جمله

... بر موضع الشمس که از جداول گرفته اند نهند

و سمارا در احد انقلاب ... فرو برند وچون زمانی

بگذرد که اوج از موضع خود حرکت کرد . باش ...

سمار رابراورد . باز همان حرکت کند ...

باب ... وم در استخراج اوساط کواکب

وقتی که خواهند اوساط کوکب در ان وقت

استخراج کند ان وقت را از تاریخ فارسی

معلوم کند اگر در سالها پی نا قصه مبسوط باشد

بعینه آنچه جرا از آن سال کوشته شد . برای هر کو ... کس ...

f.18r

1   THE SECOND TREATISE:
On the Operation of the Plate of Heavens,

2   Containing Fifteen Chapters and a Conclusion.
CHAPTER ONE:

3   On the Arrangement of the Instrument. The apogee
of the sun, which had been put on the plate,

4   put opposite the apogee of the sun on the divisions of the ring,

5   i.e. put it on the position of the sun which has
been obtained from the table.

6   And stick the peg into one of the holes, and
when a time

7   passes, so that the apogee should have moved
from its position,

8   the peg having been withdrawn, give (the plate)
that motion,

9   CHAPTER TWO: On the Extraction of the Mean
(Positions).

10   Whenever it is desired to extract the mean
(positions) of the planets at that (i.e. at
such and such a) time,

11   determine that time from the Persian calendar.

12   If it is in the explicit years (of the table),

13   take exactly what is written opposite that year
for each planet.

We have set the beginning of the mean motions at the incomplete (or explicit) year 851, which is the first of the year Yazdigerd) year of the enthronement of His Majesty the sovereign upon the seat of rulership, may God strengthen the throne of the kingdom by its endur- ing until the Day of Judg- ment, and increasing the marks of his justice and presence for all time. Amen, O Lord of the Worlds!

81

وسال طلب

گیرند و اگر بیشتر یا کمتر باشد باز آن سالها که
بین او و مقدار که از سالهای نامه مجموعه باشد
گیرند و آنچه در باز آن سال قصه است با آنچه
در باز آن سال قصه کرفته اند جمع کنند اگر سال طلوع
از سالهای ناقصه بود و آلاآنچه باز آن سالها
کرفته اند را آنچه باز آن سالهای کرفته اند نقصان
سنی و ماه و روز مطلوب از جدول
شهور و ایام آنجه یابند بر حاصل از سال افزایند
در ل و ماه و روز مطلوب حاصل شود و در
نصف النهار بطول قسطنطنیه صیانت می حفظ
والیها عن الافات والبلیة و آن سه قاست
واگر بطول ی غیر سه تا خواهد ما سن الطولین
را ساعت و دقائق کرده هر چند که شود حاصل

۹۵

f.18v

1   But if it be before or after, take opposite a
year (such) that

2   between it *and the desired year* there be to
the amount of one of (the tabulated) sum(s)
of complete years.

3   And that which is opposite that incomplete
year to that which

4   has been taken opposite that complete year,
add if the desired year

5   is after the incomplete years (tabulated).
Otherwise that which was obtained opposite the

6   complete years, and that which was obtained
opposite the incomplete years, subtract
(them).

7   Then, that which is found opposite the desired
month and day from the table of

8   months and days, add to the result (obtained
for) the first of the year

9   so that the mean (position) in the desired
year and month and day result, for

10   noon at the longitude of Constantinople, may
she preserve in safekeeping

11   her lord from calamities and affliction, and
that is 60;0°.

12   And if (it) is desired at a longitude other
than 60°, (the amount) between the two longitudes

13   having been made (into) hours and minutes,
whatever it become, the result of

---

* Marginal addition.

حرکت یکساعته را که در جدول نهاده ایم در این
ساعت و دقایق ضرب کنند و حاصل ضرب را
بر حاصل وسط در نصف النهار سیه زیاده کنند
اگر طول مطلوب کمتر باشد و نقصان کنند اگر پیشتر
بود و همچنین اگر در وقتی غیر نصف النهار
خواهند ساعات و دقایق گذشته از نصف
النهار یا باقی تا نصف النهار را در حرکت وسط
در یکساعت ضرب کنند و حاصل را بر حاصل
مذکور را فزایند اگر بعد از نصف النهار بود و نقصان
کنند اگر قبل از نصف النهار باشد حاصل وسط
مطلوب بود در وقت مطلوب و وسط
شمس به عینه وسط سفلین است و این ساوی
وسط و خاصه هر یک از علویه است یعنی

f.19r

1   the motion (for) one hour, which we have put
in the table, multiply by that (number of)

2   hours and minutes, and increase by the product

3   the result of the mean (motion) for the noon
of (longitude) 60

4   if the desired longitude is less, and decrease
(it) if (it) is more.

5   And in the same manner, if (it) is wanted at
a time other than noon,

6   the hours and minutes passed beyond noon,

7   or those remaining until noon, multiply by
the motion of the mean

8   in one hour, and the result is added to the

9   (above-)mentioned result if (it) is after
noon, and decrease (it)

10   if it is before noon, the result will be the

11   desired mean at the desired time. And the
mean of the

12   sun is exactly the mean of the two inferior
(planets), and this is equal to

13   the mean anomaly of each of the superior
(planets), i.e., it is

ساوی خاصه مرکبه ایشانست وچون وسط آفتاب

یینه وسط سفلیین است وسط ایشانرا درجدول

نهادیم ومجا وی وسط خاصه مرکبه ایشان از ابو نوشته

ایم ما بـــــسیوم در معرفت تقویم

آفتاب وتعدیل او وبعدکش ازمرکز عالم بمثل

وسط از اجـــزا جحره علامتی نهند وما را علامت

وبجاوبنیم پس حرف مطره رابر علامت

وسط و مرکز ستعار بگذرانند چنانکه مبدأ اقسام

مطره بر علامت وسط افتد وحرف عضاده را

موازی مسطره که دانند یعنی دو قوس واقع

بوزن الـــط طبین از دو طرف متساوی کرده

وبرموضع عضاده از اجـــزا جحره بغرب علامت

وسط از اجـــزا جحره علامتی نهند وآنرا موضع مقوم

f.19v

1   equal to their compound anomaly. And since
     the sun's mean

2   is exactly the mean of the inferior (planets),
     we have not put their mean (longitudes) in the
     table,

3   and in place of it we have written their
     compound anomalies.

4   CHAPTER THREE. On the Determination of the
     True Longitude of the

5   Sun, and Its Equation, and Its Distance from
     the Center of the Universe. According to
     (the sun's)

6   mean (longitude) put a mark on the divisions
     of the ring, and this we call the mark of the

7   mean. Then, make the edge of the ruler to
     pass alongside the mark of the

8   mean and the fictitious center so that the
     beginning of divisions of the

9   ruler fall on the mark of the mean. And (then)
     the edge of the alidade

10   is made parallel to the ruler, i.e., the two
     arcs which are formed

11   between the two rulers having been made equal.

12   And at the position of the alidade at the
     divisions of the ring, near the mark of the

13   mean at the divisions of the ring, a mark is
     placed and that we call the true position.

_s is in order that the
_tance of the sun from
_ center of the universe
_be ascertained.

خوانیم از اول حمل تا موضع مقوم تقویم بود و ما بین علامت

وسط و موضع مقوم تعدیل باشد و میان مرکز مستعار

و علامت وسط از جزای مسطره بعد آفتاب

از مرکز عالم با جزایی که نصف قطر خارج شصت

باشد والله اعلم باب چهارم

در معرفت تقویم ما چون وسط شمس را از وسط قمر

نقصان کنند بعد ماند و چون تضعیف بعد کننده مرکز

قمر معلوم گردد و اینرا بعد مضاعف خوانند پس مری

عضاده را بر مثل مرکز قمر از اجزا جمره به

بر موضع تقاطع حرف عضاده بمنطقه قمر علامت

کنند این علامت مرکز قمر بود پس حرف مسطره

را بر نقطه محاذاة و علامات مرکز نهند

و عضاده را موازی مسطره کنند و بر موضع قمر مری عضاده

f.20r

1   From the first (point) of Aries to the true
position will be the true longitude, and (the
angular distance) between the mark of the

2   mean and the true position is the equation.
And between the ficticious center

3   and the mark of the mean, (measured) in divisions of the ruler is the sun's distance

4   from the center of the universe, in such
divisions that half the external (i.e.,
deferent) diameter is sixty,

5   but God knows better. CHAPTER FOUR

6   On the Determination of the True Longitude
of the Moon. If the sun's mean is taken
from the moon's mean

7   a distance remains (the mean elongation), and
if the distance is doubled the center (of the
deferent) of the

8   moon becomes known, and this is called the
doubled distance (the double elongation).
Then having put the pointer

9   according to the moon's center on the divisions of the ring,

10   at the place of intersection of the edge of
the alidade with the moon's deferent, make a
mark.

11   Let this be the mark of the moon's (epicycle)
center. Then put the edge of the ruler

12   along the opposite point and the mark of the
center,

13   and make the alidade parallel the ruler, and
at the place where the pointer of the alidade
has fallen

از اجزا دهجره علامتے کنند و این مبدا حرکت
خاصه بود و بعد از این عضاده را بر خلاف توالی
حرکت دهند بمقدار خاصه قمر یعنی خاصه
هر چند برج و درجه و دقیقه که بود از این مبدا آن قدر
بر درجه بشمارند هر جا که منتهی شود بر قسم
اختلاف علامتے کنند و لا جا بر خلاف جهت
منتهی حرکت خاصه بود و این علامت اختلاف
قمر بود و بعد ذلک حرف سطر را بر هر دو علامت
مرکز و اختلاف نهند و عضاده را موازی که دانند
بر موقع مری عضاده بمقرب علامت مرکز علامت
کنند بر اجسم و دهجره و این را موضع مقوم یا بیم پس
تفصیل وسط قمر ابر هرکز بر یا بین اول و حل و موضع
مقوم بریابد و کردانند تقویم حاصل شود و الله اعلم

f.20v

1   make a mark at the divisions of the ring. And
     this is the beginning of the motion of the

2   anomaly. And after this move the alidade
     contrary to the succession (of the zodiacal
     signs)

3   to the amount of the lunar anomaly, i.e., the
     anomaly

4   however many signs and degrees and minutes it
     be, from this beginning, count off that many

5   signs and degrees. Wherever it ends, at the
     mark of the

6   difference, make a mark. And make sure it is
     opposite the direction of the

7   end of the anomalistic motion, and let this be
     the difference mark of the

8   moon. And after this, the edge of the ruler
     is placed along both the mark of the

9   center and the difference, and make the alidade
     parallel (to the ruler).

10   At the position of the pointer of the alidade
     near the mark of the center make a mark

11   on the divisions of the ring, and this we call
     the true position. Then

12   the excess of the moon's mean over the center
     between the first (point) of Aries and the

13   true position, let (it) be increased. The true
     longitude results, but God knows better.

باب پنجم در معرفت تقویم جمله نجیره

مری عضاده بر مشکل وسط گیرند از اجزاء حجره
و حرف مسطره را بر نقطه به معدل لمسیر گذرانید
هواری عضاده سازند پس بر موضع تقاطع حرف
مسطره با منطقه هر کوکبی که علامت کنند آن علامت
مرکز آن کوکب بود و پس مری رأس عضاده و
بر وسط شمس نهند بجهت هر یک از کواکب
علویه و بر رقصم اختلاف هر یک علامت کنند
کنند بر صفحه این علامت اختلاف طولی
آن کوکب بود و لامحاله علامات اختلاف
علویه بر خطی وهمی باشند که واصل بود میانه
مرکز نظر و وسط آفتاب و بجهت سفلیین مری
عضاده را بر خاصه هر که لمجریک نهند و بر برو قسم

f.21r

1   CHAPTER FIVE. On the Determination of the
    True Longitude of the Five Planets.

2   Place the pointer of the alidade according to
    the mean (longitude) on the divisions of the
    ring,

3   and, having made the edge of the ruler to pass
    alongside the equant,

4   make it parallel the alidade. Then, at the
    place of intersection of the edge of the

5   ruler with the deferent of each planet make a
    mark. This shall be the

6   center mark of that planet. Then put the
    pointer of the head of the alidade

7   at the mean of the sun in the case of each
    one of the

8   superior planets, and at the difference mark
    of each make a mark

9   on the plate. This will be the longitudinal
    difference mark of

10 that planet, and make sure that the difference
    mark (in the case) of a

11 superior (planet) is always along a line
    which joins

12 the center to the opposite (i.e., the supple-
    ment) of the sun's mean (longitude). But as
    for the inferior (planets), the pointer of the

13 alidade is put along the compound anomaly of
    each one, and at the mark of the

93

اختلاف بر صفحه علامتی نهند و این علامت
اختلاف آن کوکب بود و این نیز در جهت
بطه خاصه مرکبه واقع خواهد شد و بعد ذلک حر
سطه ابر علامت مرکز و اختلاف هر کوکب
کذرانند بر موقع مری عضاده که بقرب علامت
مرکز بود علامتی نهند و آن موضع مقوم و اذا دل
حل تا موضع مقوم تقوم آن کوکب بود و الله اعلم

باب ششم در معرفت تعدیلات و معرفت
مرکز و خاصه معدله اگر چه در استخراج تقویم با این
اثت احتیاج با این تعدیلات نداریم لکن چون خواهیم
که بدانیم بر شکل وسطه هر کوکب بر حجره علامتی کنند
و حرف عضاده را بر علامت مرکز کذرانند بر
مری عضاده بقرب وسط علامتی و بکر کنند بر حجره

f.21v

1   difference put a mark on the plate. And this is the mark of the

2   difference of that planet, and this also will befall in a direction

3   opposite to that of the compound anomaly. And after that, make the edge of the

4   ruler pass along the mark(s) of the center and the difference of each planet,

5   *and put the edge of the alidade parallel to the ruler.* At the position of the pointer of the alidade which is near the mark of the

6   center a mark is put, and that will be the true position, and from the first (point) of

7   Aries to the true position will be the true longitude of that planet,

8   but God knows better. CHAPTER SIX. On the Determination of Equations, and the Determination

9   of the Center, and the Adjusted Anomaly. Although for the extraction of the true longitudes with this

10   instrument we do not need these equations, nevertheless if we want

11   to know (them), according to the mean of each planet, a mark is put on the ring,

12   and, having put the edge of the alidade alongside the mark of the (epicycle) center, at the

13   pointer of the alidade near the mean (i.e., on the same side as the mean), make another mark. On the ring,

---

*Marginal correction.

مابین العلامتین از اجزا بقدر تعدیل شمس وتعدیل

اول بجنس بود و تعدل اول قسمه بقدر مابین

علامت ثانی و علامت مبدأ حرکت خاصه بود

وتعدیل ثانی متغیر بقدر مابین علامت ثانی و موضع

تقویم آن کوکب بود و موضع علامت ثانیه وسط

معدل شمس و هر یک از کواک محتبنی بباشد

وچون اوج هر کوکب را از وسط معدل او نقصان

کنند مرکز معدل آن کوکب باقی ماند وچون وسط

معل هر یک از علویه را از وسط شمس و وسط

معدل هر یک از سفلیه را از خاصیه مرکبه او نقصان

کنند خاصه معدل آن کوکب باقی ماند والله اعلم

باب بیست هفتم در معرفت عروض کواکب

ستة بجهت قمر وسط قمر چون بسیر دارا بر تقویم قمر آرند

f.22r

1   between the two marks (measured) in the divisions of the ring will be the equation of the
sun and the

2   first equation of the planet(s). And the first
equation of the moon will be to the amount
between

3   the second mark and the mark of the beginning
of the anomalistic motion.

4   And the second equation of a planet will be to
the amount between the second mark and the

5   true position of that planet. And the place
of the second mark, let it be the

6   adjusted mean of the sun and any one of the
planets.

7   And if the (longitude of the) apogee of each
planet be subtracted from its adjusted mean,

8   the adjusted center of that planet will remain.
And if the

9   adjusted mean of each of the superior (planets)
be subtracted from the sun's mean, and the

10   adjusted mean of each of the inferior (planets)
from its compound anomaly,

11   the adjusted anomaly of that planet remains,
but God knows better.

12   CHAPTER SEVEN. On the Determination of the
Latitudes of the Six Planets.

13   In the case of the moon, increase the mean of
the nodes by the true longitude of the moon;

حصه عرض حاصل شود و پس مری عضاده را بر مثل

حصه عرض یا بر نظیر حصه عرض نهند از اجزاء حجره

و نظر کنند که نقطه تقاطع حرف عضاده با دائره

عرض قمر بر چند خط واقع شده است و خطوط عرض

آن قدر درعرض بود و پس حصه عرض اگر کمتر از شش

برج باشد عرض او شمالی بود و اگر بیشتر از شش

برج باشد عرض او جنوبی بود و جهت عرض علویه

و عرض ثانی سفلین بر مرکز معدل مریک از کواکب

علویه ما بین اوج و راس آن کواکب بیفزایند

ما بین اوج کوکب علویه		و راس النار
زحل	مشتری	مریخ
قم ع	ع ع	صص ط

و حاصل را مرکز عرض نامیم و در سفلین مرکز

f.22v

1   the argument of the (lunar) latitude will result.
Then the pointer of the alidade is put according
to

2   the argument of the latitude, or corresponding
to the argument of the latitude on the divisions
of the ring.

3   And observe that the point of intersection of
the edge of the alidade with the circle of

4   latitude of the moon, on how many lines it has
fallen, of the latitude lines.

5   That will be the amount of the latitude. Then
the argument of the latitude, if it is less
than six

6   signs, its latitude will be northern; and if
it is more than six

7   signs its latitude will be southern. As for
the latitude of the superior (planets)

8   and the second latitude of the inferior planets,
increase by the adjusted center of each one of
the

9   superior planets between the apogee and the
ascending node of that planet,

Between the apogee of a superior (planet) and its ascending node:		
♄	♃	♂
140;0	70;0	95;0

10 (table row 1)
11 (table row 2)
12 (table row 3)

13   and the result we call the center of latitude.
And with the inferior planets, with the

معدل عمل را تمام کنیم پس هری عضاده را بر مثل

خاصّه معدله از اجزا و حجره و ثهمه و بر صفحه بر قسم

اختلاف علامتی کنند و این را علامت اوّل نامیم

پس عضاده را بر قطر استواقائم کردانند و حرف

سطر را بر علامت اوّل کذرانند و موازی عضاده

سازند و بر تقاطع حرف سطر و قطر استوا علامت

و یک کنند و این را علامت ثانیه خوانیم پس حرف عضاده

را بر قطر استوا منطبق کردانند و علامت ثانیه را بر

عضاده نقل کنند و عضاده را بقدر غایت میل قطر

مایل بذر و و حضیض هر کوکب از بعد ا اجزاء بکردند

غایت میل قطر مدرقات و حصص

زحل	مشتری	مریخ	زهره	عطارد
و ه	ل ب	مکه	کب	قله

T R A N S L A T I O N

f.23r

1 adjusted center we finish the operation. Then
the pointer of the alidade is put according to
2 the adjusted anomaly (measured) in the divisions of the ring. And on the plate by the
mark of the
3 difference a mark is made, and this we call
the first mark.
4 Then let the alidade be erected perpendicular
to the equating diameter, and, the edge of the
5 ruler having been put beside the first mark,
make it parallel the alidade.
6 And at the intersection of the edge of the
ruler and the equating diameter let
7 another mark be made, and this we call the
second mark. Then the edge of the alidade
8 is turned so that it coincides with the equating diameter, and the second mark
9 is transferred to the alidade. And the alidade
is turned by the amount of the maximum inclination of the (epicyclic) diameter
10 passing (through) the epicyclic apogee and
perigee of each planet, (measured) from the
beginning of divisions.

11	Maximum inclination of the (epicyclic) diameter passing (through) the (epicyclic) apogee and perigee				
12	♄	♃	♂	♀	☿
13	4;30°	2;30°	2;15°	(-)2;30°	6;15°

پس بر صفحه بر سوقع علامت ثانیه که بر حرف عضاده

کرده بودند علامتی از ازا علامت ثالثه نایم پس

برای علویه مری عضاده را بر خط غایت سهل

را ازمیلی نهند و حرف مسطره را بر علامت

غایت سهل ها ل کواکب از مثل					
قمر	مشتری	مریخ	زحل	ونس	عطارد
سل	ب	ا ل	ا ی	ی ع	ه ی ع

ثالثه که راند موازی عضاده سازند و بر صفحه از

علامت ثالثه بر سوقع حرف مسطره به بعد رقم

اختلاف آن کوکب هر جهت مبدا هر جزا اسی

خطی کشید و این خط را خط سهل یا مم و جهت سفلین

حرف مسطره را بر علامت ثالثه نهاده مسطره را

موازی قطر استوا کرده اند و از علامت ثالثه

f.23v

1   Then, on the plate, at the place of the second
mark, which had been made on the edge of the
alidade,

2   a mark is made, and that we call the third
mark. Then,

3   for the superior (planets), the pointer of the
alidade is put according to the maximum inclination

4   of the inclining (deferent plane) from the
parecliptic, and, the edge of the ruler having
been placed along the

5	Maximum inclination of the inclining of the planets from the parecliptic.					
6	♄	♃	♂	♀	☿	☾
7	2;30°	1;30°	1;0°	0;10°	0;45°	

8   third mark, it is made parallel to the alidade.
And on the plate, from

9   the third mark on the place of the edge of
the ruler to an amount (being) the

10   difference mark of that planet in the direction of the beginning of divisions in proper
style

11   a line is drawn. And this line we call the
line of inclination. And for the inferior
(planets)

12   the edge of the ruler having been put along
the third mark, the ruler is made to

13   parallel the equating diameter, and from the
third mark

برصفحه درجهت مبدأ اجزا بریدد بعد رقم اختلاف آن
کوکب خط مایل کشید پس برنقطهٔ استوا بعرب
نقطه عرض علامتی کند که بعد ازآن علامت ثانیه
بقدر بعد علامت ثالثه باشد ارنقطهٔ عرض بعید
اگر مرکز عرض اقل ازشش برج بود والا ازنقطهٔ
عرض قریبه باستعمالهٔ اجزا مسطره یا ابرکار
واین علامت را بدل نقطه عرض نامیم و برخط ملل
نقطه طلب کنیم که بعد میان آن نقطه و بدل نقطهٔ
عرض مساوی بعد میان بدل نقطه عرض علامت
اولی باشد باستعمالهٔ برکار یا مسطرهٔ واین ازنقطهٔ
مطلوبه خوانیم بعد ازان حرف بسطه را ازنقطهٔ
مطلوبه و بدل نقطه عرض که رابآنجا نگاهدارند وعضو
را موازنی بسطه کردانند وعضا و مراموازی بسطه

f.24r

1   on the plate, in the direction of the beginning
of divisions, to the amount of the difference
mark of that

2   planet, the line of inclination is drawn. Then,
on the equating diameter near

3   the point of latitude a mark is made such that
its distance from the second mark shall be

4   to the amount of the distance of the third mark
from the far latitude point

5   if the center of latitude is less than six
signs, and otherwise from the near

6   latitude point with the assistance of the divisions of the ruler or compass,

7   and this mark we call the substitute for the
point of latitude. And on the line of inclination

8   we seek a point such that the distance between
that point and the substitute for the point of

9   latitude shall be equal to the distance between
the substitute for the point of latitude [and]*
the

10   first mark, with the help of a compass or ruler,
and we call this the

11   desired point. After that, the edge of the
ruler having been put along the

12   desired point and the substitute for the point
of latitude, hold it (there), and let the alidade be made

13   parallel to the ruler. And, the alidade being
parallel to the ruler,

---

*Insert ﺝ here.

وحینئذٍ نظر کنند بر مری عضاده که بر کدام کسر
افتاده است از اجزاء حجر . و بعد او از طرف
قطر استوا جنداست این غایت میل حجر کوکب
از آنرا وی بر بود از سطح مایل در سمت یمین و در علو یمان
اجزا را از غایت میل مایل از سمت نقصان کنند
اگر خاصه بعد از اقل از دیج یا اکثر از ثلثه او یا ربع
بود والا از زیاده کند تا غایت میل حجر کوکب
از بعد و پریز از سطح بر وج حاصل شود پس عضاده را
بر قطر استوا قایم کرد اینده بر حرف عضاده بر نوع
تقاطع حرف عضاده یا خطی از خطوط ما عرض
مساوی غایت میل حجر کوکب از آنند و بر
اسطح میل بر وج اعلا می کنند و ما آنرا علامت
عرض خوانیم و بعد از لک مری عضاده را به

f.24v

1   thereupon observe along (to) the pointer of the
alidade (to see) on which division it

2   has fallen of the divisions of the ring, and
its distance from the direction of the equating
diameter

3   is how much. This will be the maximum [incli-
nation]* of the part of the planet

4   from the epicycle from the inclined surface,
(i.e. deferent plane) in the inferior (planets);
and in the superior (planets), those

5   parts of the extremity of the inclination of
the deferent, let (them) be decreased from the
parecliptic

6   if the adjusted anomaly is less than a quadrant
or more than three quadrants,

7   and otherwise (it) is (to be) increased until
the maximum inclination of the part of the
planet

8   from the epicycle from the plane of the signs
(i.e., the ecliptic) results. Then, the alidade
having been erected

9   perpendicular to the equating diameter, along
the edge of the alidade, at the place of

10   intersection of the alidade edge with (that)
line of the latitude lines which is

11   equal to the maximum inclination of the part
of the planet from the epicycle

12   to the surface which is inclined with (respect
to that of the) signs, let a mark be made
(there). And we call this the mark of

13   latitude. And after that, the pointer of the
alidade having been placed

---

*Read ميل for مثل .

مثل مرکز عرض از اجزا جحر بنهاده بکار کنند
که علامت عرض بر که ام خط واقع شد و این از
خطوط عرض آن عرض مطلوب بود پس که مرکز
عرض علویه که از نشش برج بود و عرض شمالی باشد
و اگر بیشتر باشد جنوبی بود و اگر مرکز معدل
سفلیین کمتر از شش برج و حاصه معدل کمتر
از سه برج و یا بیشتر از نه برج بود و یا مرکز معدل
بیشتر از شش برج و حاصه معدل بیشتر از سه برج
و کمتر از نه برج بود و عرضش شمالی زهره شمالی
و ازان عطارد جنوبی بود و الا ازان رهس جنوبی
و عطارد شمالی باشد العرض ثالث سفلیین
نقد بل نمای کوکب را در بعد ابعد بدابند و اگر مرکز
معدل کمتر از سه برج یا بیشتر از نه برج باشد

f.25r

1   according to the center of latitude on the
     divisions of the ring, note
2   that the latitude has fallen on which line of
3   the lines of latitude; this will be the desired
     latitude. Then, if the center of
4   latitude of the superior (planet) is less than
     six signs its latitude will be north,
5   and if (it) is more, (its latitude will) be
     south. And if the adjusted center of the
6   inferior (planet) is less than six signs and
     the adjusted anomaly less than
7   three signs or more than nine signs, or less
     the adjusted center
8   more than six signs and the adjusted anomaly
     more than three signs
9   and less than nine signs, the second latitude
     of Venus will be north,
10   and that of Mercury south; and otherwise that
     of Venus south
11   and Mercury north. But (as for) the third
     latitude of the inferior (planets), let the
12   second equation of the planet be known at the
     greatest distance (i.e. at the apogee), if
     the
13   adjusted center is less than three signs or
     more than nine signs,

والا تعدیل مائی کوکب را در مقابلهٔ اوج بدانند

و ثلث سدس آن تعدیل را بگیرند در زهره

و نجهت عطارد و تعدل را هفت و قیقهٔ ضرب

کنند اگر در بعد ابعد کرفت باشند والا در هشت

و قیقهٔ ضرب کنند انحراف حاصل کرد و پس

مثل انحراف از خطوط عرض بجویند و عضاده را

بر خط استوا قائم کرد ایند و نظر کنند که خط

مطلوب یعنی که خطی که ساوی انحراف

بر کدام جزو واقع شده است اذا اجزا حرف

عضاده بران جبز علامتی کنند این علامت

عرض بود پس نو در جب بر مرکز معدل هر کوکب

زیاد کنند و حاصل را مرکز عرض خوانیم پس

مری عضاده را بر مرکز عرض نهند بار نظر مرکز عرض

f.25v

1   and otherwise let the second equation of the
planet be determined according to the apogee,

2   and take a third of a sixth of the equation
for Venus,

3   and for Mercury, multiply the equation by seven
minutes

4   if (it) is obtained on (the side of) the apogee,
otherwise multiply it by eight

5   minutes; the obliquity results. Then

6   seek the lines of latitude according to the
obliquity, and, the alidade

7   having been erected perpendicular to the equating line, note that the

8   desired line, that is, that line which is equal
to the obliquity,

9   on which division had (it) fallen of the divisions of the edge

10   of the alidade. On that division make a mark.
Let this be the mark

11   of latitude. Then let ninety degrees be increased
by the adjusted center of each planet,

12   and the result we call the center of latitude.
Then let the

13   pointer of the alidade be placed along the center
of latitude, or opposite the center of latitude

نهند از اجزاء حجره و نظر کنند که علامت عرض
بر کدام خط واقع شد . است از خطوط عرض عرض
ثالث آن کوکب بود . پس اگر مرکز معدل آن کوکب
از سه برج کمتر و یا از نه برج بیشتر باشد و خاصه
معدله اگر شش برج بیشتر باشد عرض ثالث
زهره جنوبی و عطارد شمالی والا زهره شمالی و
عطارد جنوبی بود اما عرض اول سفلیین مری
عضاده را بر مرکز عرض یا بر نظیر مرکز عرض
نهند از اجزاء حجره و نظر کنند که حرف عضاده
با دائرهٔ عرض ان کوکب بر کدام خط از خطوط
عرض تقاطع کرد . است بر آن خط علامتی کنند
و عضاده را بر قطر استوا قایم گردانند و نظر
کنند بر موضع تقاطع حرف عضاده . با خط معلم

f.26r

1   place it, from the divisions of the ring, and
     observe that the latitude mark

2   on which line of the latitude lines it has
     fallen, (this) will be the

3   third latitude of that planet. Then, if the
     adjusted center of that planet is

4   less than three signs, or more than nine signs,
     and the

5   adjusted anomaly more than six signs, the
     third latitude of

6   Venus will be south and (that of) Mercury north,
     and otherwise Venus will be north and

7   Mercury south. But (as for) the first latitude
     of the inferior (planets), let the pointer of
     the

8   alidade be put along the center of latitude,
     or opposite the center of latitude,

9   from the divisions of the ring, and observe
     that the edge of the alidade

10   has intersected with the latitude circle of
     that planet at which line of the lines of

11   latitude. On that line let a mark be made,

12   and, the alidade having been turned perpendi-
     cular to the equating diameter, observe

13   the place of intersection of the edge of the
     alidade with the marked line,

و بران موضع از حرف عضاده علامتی کنند

این علامت عرض بود پس مری عضاده را اعا

کنند بر مر که عرض یا بر نظیر مر که عرض علامت

عرض بر کدام خط که واقع است از خطوط عرض

عرض اول آن کوکب بود و عرض اول از هر

دایماشمالی و عطارد و جنوبی باشد پس چون

عرض تکمله هر کوکب را معلوم کرده باشد

اگر همه در جهت موافق باشند جمع کنند و الا عرضها

موافق راجع کرد تفاضل من المجموع و المخالف

بکیه بذ عرض معدل حاصل شود و جهت عرض معدل

جهت مجموع بود در اول و جهت فضل

بود در ثانی و الله اعلم باب هشتم

در معرفت ابعاد کواکب از مرکز عالم بد آنکه بعد

f.26v

1   and at that place on the edge of the alidade
    make a mark.

2   Let this be the latitude mark. Then let the
    pointer of the alidade be returned

3   along(side) the center of latitude, or along
    the opposite to the center of latitude, the
    mark

4   of latitude, on whichever line of the latitude
    lines (it) falls will be

5   the first latitude of that planet. The first
    latitude of Venus is

6   always north, and Mercury south. Then, since
    the

7   three latitudes of each planet have been found,

8   if all are in a similar direction, let them be
    added, and otherwise,

9   like latitudes having been added, the difference
    between the sum and the opposing (latitude)

10 is obtained '(that) the adjusted latitude result,
    and the direction of the true latitude is the

11 direction of the sum first and the direction
    of the difference

12 secondly, but God knows better. CHAPTER EIGHT

13 On the Determination of the Distances of the
    Planets from the Center of the Universe. Be
    it known that the distance of the

علامت مرکز هر کوکب از علامت اختلاف
آن کوکب مساوی بعد مرکز جرم آن کوکب است
از مرکز عالم و بعد علامت مرکز هر کوکب از مرکز
صفیحه بقدر بعد مرکز تدویر آن کوکب است
از مرکز عالم و این ابعاد معلوم است با جزائی که
نصف قطر صفیحه شصت جزء باشد از قبل سطرین
و عادت قابل این صناعت بر آن جاری شده
است که تقدیر ابعاد کواکب متحیره و شمس با جزائی
کنند که نصف قطر حوامل ایشان شصت باشد
و تقدیر ابعاد قمر با جزائی که نصف قطر مایل
شصت باشد پس اگر خواهند که از این اجزاء
معلومه آن اجزاء را که نقطه بر بان می کنند
بدانند هر یک از این اجزاء معلومه را در مقدار نصف قطر

f.27r

1   mark of the center of each planet from the
    difference mark of

2   that planet is equal to the distance of the
    center of the body of that planet

3   from the center of the universe, and the distance
    of the center mark of each planet from the center
    of the

4   plate is (equal) to the amount of the distance
    of the epicycle center of that planet

5   from the center of the universe. And these
    distances are determined in such divisions
    that

6   half the plate diameter is sixty divisions from
    the sides of the two rulers.

7   And the custom of those who are of this art has
    thus become current,

8   that the evaluation of the distances of the
    planets and the sun is made with divisions

9   such that half the diameter(s) of their deferents
    shall be sixty.

10 And the evaluation of the distances of the moon
    are with divisions such that half the inclined
    diameter

11 is sixty. Then, if it is desired that from
    these

12 apparent divisions (the number of) those divi-
    sions which (are used to) evaluate them

13 be known, multiply each of these (values in)
    known divisions by a quantity

کننده که نسبت آن مقدار با یک جزو از اجزاء
قطر یعنی نسبت نصف قطر صفیحه بود یا نصف
قطر منطقه کامل آن کوکب و ما این مقدار در هر
کوکب استخراج کردیم در جدولی نهادیم و آن
جدول اینست

۴۴۱	۴۴۱	۱ الطا ۵ که ۱	الطوس ۱۱	ح ۱	خو ۶

و چون نصف قطر صفیحه بمقدار نصف قطر
خارج مرکز شمس است به محیط صفیحه منطقه
خارج مرکز شمس است کما فی المقاله الاولی
و بمقدار نصف قطر مایل است ابعاد معلومه قبل
مسطرین همان بعد بمقدار مطلوب بود والله اعلم

T R A N S L A T I O N

f.27v

1 such that the ratio of this quantity to one division of the divisions of the
2 diameter be the same as the ratio of half the plate diameter to half the
3 diameter of the deferent heaven of that planet. And we, having extracted this quantity for each
4 planet, have put it in a table, and that
5 table is this:

6	☉	☾	♄	♃	♂	♀	☿
7	1;0,0	1;0,0	1;9,11	1;4,55	1;19,12	1;1,3	1;13,56

8 And since half the plate diameter is (equal) to the amount of half the diameter of the
9 eccentric (orbit) of the sun, thus the plate circumference is the
10 eccentric deferent of the sun, as was prescribed in the First Treatise,
11 and (it) is to the amount of half the diameter of the inclined (orbit). The apparent distance, ascertained by means of the
12 two rulers, that same distance will be the desired quantity, but God knows better.

باب نهم در معرفت رجعت و
استقامت و اقامت کواکب و اوقات آن
برای معرفت رجعت و استقامت و اقامت
تقویم کواکب را در ایام متوالیه حاصل کنند اگر
متزاید باشد مستقیم بود و اگر متناقص بود راجع
باشد و اگر نه متزاید و نه متناقص بود و مقیم باشد
اما معرفت مقامات رجعت و استقامت
وقتی که خاصه سعد لبعد و مقامات رجعت
و استقامت که در جدول موضوع است برسد
بعد مرکز تدویر را از مرکز عالم به جزء اول صفحه
بدانند و آن بعد علامت مرکز است از مرکز صفحه
و از ابعاد حفوظ غرایم بعد از آن به رابع و بعد
اقرب مرکز تدویر آن کوکب را به جزء دانند

f.28r

1   CHAPTER NINE. On the Determination of Retro-
     gradations

2   and Forward (Motions) and Stations of the Pla-
     nets and their Times.

3   For the determination of retrogradations, and
     forward motions, and stations,

4   extract the true longitude of the planet for
     successive days. If

5   it (the true longitude) is increasing it (the
     planet) will be (in) direct (motion), and if
     it is decreasing it will be retrograde,

6   and if it is neither increasing nor decreasing
     it will be stationary.

7   But, as for the determination of direct and
     retrograde stations,

8   when the adjusted anomaly reaches the limits
     of the retrograde

9   and direct stations, which are placed in the
     table, ascertain the

10   distance of the center of the epicycle from
     the center of the universe, in the divisions
     of the diameter of the plate.

11   And this is the distance of the mark of the
     center from the plate center,

12   and we call this the preserved distance. After
     that, the greatest distance and the

13   least distance of the epicycle center of that
     planet are determined in divisions of the
     diameter

وطریق معرفت آن جنانست که عضاده را بهط
محاذات که ذار انند تا با وج و حضیض گذرد بعد
نقطۀ تقاطع عضاده با منطقۀ آن کوکب از مرکز
صفحه در جانب اوج بعد ابعد بود و جانب حضیض
بعد اقرب باشد مگر در عطارد که بعد اقرب او ابعد
نیست بلکه نقطه است از منطقه اش و ترتیب
اوج و تفاضل میان بعد ابعد و بعد اقرب کواکب
و تفاضل بنهار ایا جز اینست که نصف قطر صفحه
شصت باشد در جدول نهادیم که عند الحاجة

به اینجا و پیدا بود ابعد و اقرب

ازانجا ... ار اند و عمل محتاج نشوند و آن جدول اینست

اسم	بعد ...	ریکح	ابعح	د هح	بعح	ریکح	ا
بعد ابعد	ه	ه	یب	ن	ع	سه	لوند
بعد اقرب	و	نو نو	نز	م نو	ش نو	نط ط	و
فضل	ند	و ی	ط و	نو	ه	ه نو	

f.28v

1   The method of determining that is thus: that
     the alidade is turned

2   opposite until it passes through the apogee
     and perigee. The distance of the

3   point of intersection of the alidade with the
     deferent of that planet from the center of the

4   plate on the side of the apogee will be the
     farthest distance, and on the side of the
     perigee is the

5   nearest distance, except for Mercury, for
     which the nearest distance

6   is not this, but it is a point of its deferent
     (found by) dividing (its deferent) into three
     parts (from the)

7   apogee. And the differences between the
     farthest distance and the nearest distance
     *is obtained. And the greatest and least
     distances* of the planets

8   and the differences between the two in divi-
     sions such that half the diameter of the plate
     shall be

9   sixty, we have put in a table that, when
     needed,

10   it may be taken from there. And that table
     is this:

	Distances	♄	♃	♂	♀	☿	
11	Distances	♄	♃	♂	♀	☿	
12	Farthest distance	55;0	58;0	50;0	60;0	56;0	Annulled
13	Nearest distance	49;4	52;56	40;54	57;56	45;6	Annulled
14	Difference	5;56	5;4	9;6	2;4	10;54	Annulled

---

*Marginal correction.

123

پس بعد محفوظ را اربعه ابعد نقصان کنند در علویه و
غیره و بعد اقرب عطارد را از ابعد محفوظ نقصان
کنند و باقی را در حد ما بین مبداء حدود مقامات و منتها
ضرب کنند و آن در جدول موضوع است و حاصل
ضرب را بر تفاضل من البعدین قسمت کنند و خارج قسمت را

الکواکب	ه	ه	ه	ه	ه
مبداء حدود و مقامات حصص					
منتهای حدود و مقامات رجعت					
تفاضل فصل					

بر مبداء حدود و مقامات رجعت زیاده کنند
تا مقام رجعت حاصل شود و مقام رجعت را از دور
نقصان کنند مقام استقامت باقی ماند یا خارج
قسمت را از منتهای حدود و مقامات استقامت

f.29r

1   Then subtract the preserved distance from the
farthest distance in (the case of) the superior
(planets) and

2   Venus, and the nearest distance of Mercury,
subtract it from the preserved distance.

3   And the remainder multiply between the beginning of the limits of the stations and their
ends.

4   And that is placed in the table. And the

5   product is to be divided by the difference
between the two distances, and the quotient,

	The Planets	♄	♃	♂	♀	☿
6						
7	Beginning of the limits of the retrograde stations and all the beginning of the limits of the direct (stations).	$3^s 22;45°$	$4^s 7;6°$	$5^s 7;14°$	$5^s 15;45°$	$4^s 24;29°$
8	End of the limits of the retrograde stations.	$3^s 25;29°$	$4^s 10;11°$	$5^s 18;48°$	$5^s 18;27°$	$4^s 27;14°$
9	The difference between the two.	$2;44°$	$3;5°$	$11;34°$	$2;42°$	$2;45°$

10   increase by the beginning of the limits of the
retrograde stations,

11   so that the retrograde station result. And if
the retrograde station is subtracted from a
revolution;

12   the direct station will remain. Or (else)
the

13   quotient is subtracted from the ends of the
limits of the direct station

نقصان کنند یا مقام استقامت حاصل شود مقام
استقامت باشد و در نقصان کنند مقام رجعت باقی
ماند و مقام رجعت یا مقام اول و مقام استقامت را
مقام ثانی خوانند پس هر گاه که خاصه معدله از مقام
اول بگذرد و کوکب راجع شود و چون از مقام ثانی
بگذرد مستقیم شود و اگر در مقام اول بود بعقب
رجعت بود و در مقام ثانی بعقب استقامت باشد
و ہمین را اخیر از استقامت بنو و اما که کوکب مستقیم
بود و خواہیم بدانیم که کی راجع می شود و خاصه معدله
را از مقام اول نقصان کنیم و باقی را بر حرکت
یکروز خاصه قسمت کنند خارج قسمت مدتی باشد
که بعد از ان مدت راجع خواهد شدن و اگر راجع بود
و وقت استقامتش خواہند خاصه معدله را از مقام

f.29v

1   so that the direct station result. The

2   direct station is subtracted from a revolution;
the retrograde station will remain.

3   And the retrograde station is called the first
station; the direct station the

4   second station. So, whenever the adjusted
anomaly passes the station,

5   the planet will become retrograde, and if it
passes the second station

6   it will become direct. And if it is in the
first station its stance will be

7   retrograde, and in the second station its stance
will be direct.

8   But the two luminaries (i.e. the sun and the
moon) will never be (in) other than direct
(motion). However, if a planet is direct

9   and we want to know when it will become retrograde, the adjusted anomaly is subtracted

10   from the (value at) first station, and the
remainder is divided by the

11   diurnal motion of the anomaly. The quotient
will be such a time

12   that after that time the retrogradation will
occur. And if it is retrograde

13   and the time until it becomes direct is required,
the adjusted anomaly having been subtracted
from the (value at)

ثانی نقصان کرده ... همان عمل کنند والله اعلم

باب دهم در معرفت نطاقات

اوجی وند وبری آفتاب و دیگر کواکب که در افلاک
خارج مرکز حماز نطاق بود مبدا نطاق اول اوج
باشد و مبدا نطاق سیم حضیض و اما مبدا دوم
و چهارم اگر حسب بعد کره ندا بود که بعد آفتاب
با مرکز تندوبر از مرکز عالم و مرکز خارج مرکز متساوی
بود و اگر حسب سیر که ندا انجا بود که سیر آفتاب
نه سریع بود و نه بطی و غیر آفتاب را از سیارات
در فلک تندور هم حماز نطاق بود مبدا اول وسیم
ازوج و حضیض مرای کو مبدا دوم و چهارم حسب
عدا انجا بود که بعد کوکب و مرکز تندوبر از مرکز عالم
متساوی باشد و حسب سیر انجا بود که سیر کشف سر مرکز

TRANSLATION

f.30r

1   second station, perform the same operation,
     but God knows better.

2   CHAPTER TEN. On the Knowledge of

3   Apogee and Epicycle Sectors. The sun and the
     other planets in the

4   eccentric orbits have four sectors. The
     beginning of the first sector is the apogee,

5   and the beginning of the third sector is the
     perigee. But (as for) the beginning of the
     second sector

6   and the fourth (sector), if it is (reckoned)
     on the basis of distance it will be there
     where the distance(s) of the sun

7   or of the epicycle center from the center of
     the universe and from the eccentric center
     are equal.

8   But if it is (reckoned) on the basis of move-
     ment, it will be there where the sun's movement
     would

9   not be speedy and not slow. And other than
     the sun, of the planets,

10   in the epicycle there are also four sectors.
     The beginning of the first and third

11   will be the apparent epicyclic apogee and
     perigee, and the beginning of the second and
     fourth, according to

12   distance will be there where the distance of
     the planet and the center of the epicycle from
     the center of the universe

13   are equal. And according to movement (it)
     will be there where the movement, according
     to the center,

129

تنها باشند و اين مهرو و باختلاف بعد هر کزبذ وبر از مرکز
عالم مختلف شوند و ماباری نظاقات او جی زانحب
بعد بجهت شهرتش از بح اخذ کرد . برمنطقه هر کوکب
علامات کرد يم س نطاقات او جی را از قبل علامت
بدانند و اما نطاقات تد ويری راچون برصنفه علامت
مرکز و اختلاف طولی کرد . باشند حرف عضاد
يا بر علامت مرکز نهند و نظر کند بعلامت احتلاف
اکبر بر مين کسی که مواجه عضاد . باشد بحیثیتی که
علامت مرکز محاذی را سا آن کس بود و واقع و چنین
اکر بعد علامت مرکز از علامت اکثر بود و از بعد علامت
مرکز از مرکز صنفه کوکب نطاق اول باشد و اکر اقل
باشد در ثانی بود و اکر علامت اختلاف طولی از
يسار آن کس باشد و بعد علامت مرکز از علامت

f.30v

1   is alone (i.e., the only one). And both of
    these, because of the difference in distance
    of the epicycle center from the center of the

2   universe, differ. And we, having taken from
    the (Khāqānī) Zīj the beginnings of the apogee
    sectors (reckoned) according to

3   distance, in order that they be known, we made
    on the deferent of every planet

4   some marks. So the apogee sectors can be
    determined by means of the marks.

5   As for the epicyclic sectors, if on the plate
    a mark of the

6   center and of the longitudinal difference had
    been made, the edge of the alidade

7   is put along the mark of the center, and look
    along to the difference mark.

8   If it is on the right of someone who is facing
    the alidade under the condition that

9   the mark of the center fall opposite the head
    of that person, and hence

10 if the distance of the center mark from the
    [difference]* mark is larger than the distance
    of the

11 center mark from the center of the plate, the
    planet must be in the first sector, and if it
    is less,

12 it will be in the second. And if the mark of
    longitudinal difference is on the

13 left of that person, and the distance of the
    center mark from the mark of the

---

*The word اختلاف is lacking in the text.

131

اختلاف کم از بعد علامت مرکز از هر که صحیحه بود در
نطاق ثالث بود و اگر بیشتر باشد در نطاق رابع
باشد و همگنا که کوکب در نطاق اول و دوم باشد
زاید بود در عدد حساب و در نطاق سیم و چهارم
ناقص بود و حال قمر بر عکس این باشد در زیادت
و نقصان حساب و اگر کوکب در نطاق اول و
چهارم باشد ناقص بود در عظم و نور و اگر در دوم
و سوم باشد زاید باشد اما زیادت و نقصان نور
ماه بحسب بعد او باشد از آفتاب لکن در عظم
مانند دیکر کواکب بود والله اعلم باب

یازدهم در معرفت خسوف ماه هر استقبال حقیقی
که در شب باشد یا درد و طرف روز واقع شود
کم از ده و ساعت و چهار و قیقه کذشته اذا اول روز

f.31r

1   difference is less than the distance of the
center mark from the center of the plate, it
will be

2   in the third sector, and if it is more it will
be in the fourth sector.

3   And whenever the planet is in the first and
second sectors it will be

4   increasing in the number of the computation,
and in the third and fourth sectors it will be

5   decreasing. But the condition of the moon is
opposite this in the increase

6   and decrease of the computation. And if the
planet is in the first or

7   fourth sector it will be decreased in magnitude
and light, but if it is in the second

8   or third it must be increased. But as for the
increase and lessening in light of the

9   moon, it will be according to its distance from
the sun, but in magnitude

10   it is like the other planets, but God knows
better. CHAPTER

11   ELEVEN. On the Determination of Lunar Eclipses.
(At) any true opposition

12   which is in the night or which falls on (either
of) the two sides of the day,

13   less than two hours and four minutes having
passed from the first of the day

یا مانده از آن خسه روز و عرض قمر کمتر از
مقدار مجموع قطر ماه و نصف قطر دایره ظل
بود و خسوف ممکن بود و چو اگر مقدار مجموع نصف
قطرین باشد قمر ماس دایره ظل شود و منخسف
نگردد و اگر زیاده بر مجموع نصف قطرین بود
خسوف ممکن نباشد پس اگر بعد از احدی
العقدتین اکثر از دوازده درجه باشد خسوف
ممکن نخواهد بود و اگر کمتر از دوازده درجه و بیشتر
از بیست و نه دقیقه باشد بعضی منخسف
گردد و اگر کمتر از این باشد خسوف کلی واقع شود
پس هر گاه که معلوم شود که خسوف واقع خواهد شد
حرف عشاره را باول حمل نهد و دقایق عرض
قمر ابدرجات رفخ کنند و بمقدار این درجات

f.31v

1   or remaining from the end of the day, and the
moon's latitude being less than the

2   amount of the sum of [half]* the diameter of
the moon and half the diameter of the shadow
circle,

3   a lunar eclipse is possible. Otherwise, if
the amount of the sum is half the

4   two diameters the moon will be tangent to the
shadow circle, and a lunar eclipse

5   will not occur. And if it is greater than the
sum of half the two diameters a

6   lunar eclipse will not be possible. So if
the distance of the moon from one of the

7   two nodes is greater than twelve degrees a
lunar eclipse

8   will not be possible, but if it is less than
twelve degrees and [the lunar latitude at
conjunction]* more

9   than twenty-nine minutes a partial lunar eclipse
will

10   occur, and if it is less than this a total
lunar eclipse will befall.

11   So, any time it become known that a lunar
eclipse is to befall, put the

12   edge of the alidade along the first of Aries
and elevate the minutes of the moon's latitude

13   into degrees, and to the amount of these

---

*Lacking in the text.

مجموعه از اجزاء قطر بر صفحه علامتی کنند

و مری عضاده را حرکت دهند تا حد انقلاب بین آید

و باید که حرف عضاده بجانب علامت باشد

و حرف مسطره را بربین علامت نهند بخشیتی

که علامات خسوف که بر مسطره است بربن علامت

واقع شود و بر اسرع یک سطره که قرب بعلامت

ممکن است بحرف عضاده افتد لابد از حرف

سطره و عضاده را و یه حادث شود پس

نظر کنند که میان محد و زاویه و مرکز عضاده چند

است از اجزاء عضاده اینچه باشد تضعیف کنند

پس به یک مرتبه منط کیم زند یعنی هر دو جزو را دقیقه

شمارند اینچه حاصل شود مدت ساعات سقوط بود و اگر

خسوف کلی باشد اینچه بعلامت خسوف

f.32r

1   elevated degrees in divisions of the diameter
     on the plate make a mark.

2   And the pointer of the alidade is moved until
     it reaches one of the two solstices,

3   and it is necessary that the edge of the alidade
     be beside the mark.

4   And the edge of the ruler is put alongside this
     mark in such fashion that

5   the lunar eclipse mark, which is on the ruler,
     fall on this mark,

6   and the other head of the ruler, which is near
     the mark of

7   first totality, fall on the edge of the alidade.
     Without fail, from the edge of the

8   ruler and the alidade an angle is formed. So

9   observe that between the vertex of the angle
     and the center of the alidade how many

10   of the divisions of the alidade there are. What-
     ever it is, double it,

11   then depress it one (sexagesimal) place, i.e.,
     count each degree a minute.

12   That which results is the time of immersion.
     And if the lunar

13   eclipse is total, that which was done with the
     lunar eclipse mark

کرده شد بعلامت کش کنند یا ساعات
کش معلوم کرد و پس ساعات استقبال
را در پنج موضع هند از اول ساعات سقوط و
از ثانی ساعات کش را بکاسته و ثالث
را بجایش بگذارند و بر رابع ساعات کش
و بر خامس ساعات سقوط را زیاده کنند تا از اول
ساعات بدو خسوف و از ثانی بدو کش
و از ثالث وسط خسوف و از رابع بدو انجلا
و از خامس تمام انجلا حاصل کرده و اگر خسوف
کلی بود ساعات استقبال را در پنج
موضع گذارند حاصل از اول ساعات بدو
خسوف و از ثانی وسط خسوف و از ثالث
تمام خسوف بابند سپس میل مرفوع عرض قمر

f.32v

1   do it with the mark of (first) totality [so
     that]* the hours of

2   totality can be ascertained. Then the time of
     opposition,

3   put it in five places: from the first subtract
     the time of immersion and

4   from the second the time of totality, and the
     third, put it in

5   its (previous?) place, to the fourth add the
     time of totality,

6   and to the fifth the time of immersion, so
     that from the first the

7   time of beginning of the lunar eclipse, and
     from the second the beginning of totality,

8   and from the third the middle of the lunar
     eclipse, and from the fourth the beginning
     of clearance,

9   and from the fifth the end of clearance result.
     But if the lunar eclipse

10   is not total the time of opposition is put
     in three

11   places. The result from the first will be the
     time of the beginning of the lunar

12   eclipse, and from the second the middle of the
     lunar eclipse, and from the third the

13   end of the lunar eclipse. Then, according to
     the elevate of the lunar latitude,

---

*MS. has ڶ for ڬ .

از اجزاء مسطره ازان راس که بغرب علامت
مکث است مش قسمت اصابع خسوف جوند
وازان وجه مسطره اصابع منخسفه قمر ابدانند
والله اعلم باب دوازد هم

در معرفت کسوف افتاب اگر اجتماع
در روز یا در یکی از دو طرف شب واقع شود
وجزء اجتماع بعد از راس و پیش از ذنب
وبعد ان جزء از عقده کمتر از شانزده درجه
باشد یا پیش از راس و بعد ان جزء از عقده کمتر
از شانزده درجه باشد یا پیش از راس
وبعد از ذنب باشد وبعد جزء از عقده کم از هفت
درجه بود و کسوف ممکن باشد پس هر گاه
که کسوف ممکن باشد باز ار جزء اجتماع و ساعا

f.33r

1 in divisions of the ruler, from that head which
is in the vicinity of the mark of

2 totality, in front of the division, the digits
are sought,

3 and from that face of the ruler the lunar
eclipse digits are ascertained,

4 but God knows better. CHAPTER TWELVE.

5 On the Determination of Solar Eclipses. If a
conjunction befalls

6 during the day or on one of the two sides of
the night,

7 and the part of the conjunction (is) after
the head (i.e. ascending node) and before the
tail (i.e. descending node),

8 and the distance of that part from the node is
less than sixteen degrees;

9 or before the head, and after {that part from
the node (is) less

10 than sixteen degrees; or before the head

11 and after}* the tail and the distance of the
part from the node is less than seven

12 degrees, a solar eclipse is possible. So,
whenever a solar

13 eclipse is possible, opposite the part of the
conjunction and the hours of the

---

*Passage in braces is repeated also in the text.

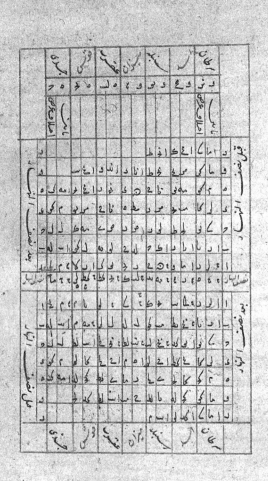

f.33v

(Table of Adjusted Lunar Parallax Components)

	♋		♌		♍		♎		♏		♐		♑		
	6;57		6;48		6;16		6;0		5;[44]		5;[2]		5;3		
(Hours) between	Latitude Difference												(Hours) between	Latitude Difference	
7	1,41	28	1,48	20	1,53	9									7
6	1,41	27	1,47	18	1,53	9	1,51	4	1,54	6	1,48	12			6
5	1,40	26	1,45	16	1,51	8	1,50	4	1,53	7	1,48	13	1,45	22	5
P.M. 4	1,37	21	1,42	13	1,47	7	1,49	5	1,5,	9	1,47	16	1,~	26	4 A.M.
3	1,28	16	1,33	9	1,36	6	1,47	7	1,47	10	1,45	20	1,30	30	3
2	1,7	11	1,11	7	1,20	6	1,34	8	1,36	15	1,30	26	1,12	35	2
1	0,37	7	0,41	6	0,50	8	1,30	13	1,16	22	1,7	31	0,40	38	1
Noon	0,0	5	0,7	8	0,15	14	0,32	20	0,13	29	0,29	37	0,40	41	Noon
1	[0]37	7	0,28	12	0,13	20	0,0	28	0,0	36	0,11	40	0,40	38	1
2	1,7	11	0,58	19	0,42	26	0,30	35	0,30	37	0,45	40	1,12	35	2
3	1,28	16	1,16	24	0,58	32	0,52	39	0,53	38	1,12	39	1,30	30	3
A.M. 4	1,37	21	1,18	29	1,8	36	1,5	40	1,8	38	1,21	36	1,46	26	4 P.M
5	1,40	23	1,21	33	1,10	38	1,7	41	1,10	39	1,23	34	1,43	22	5
6	1,41	26	1,23	35	1,11	39	1,8	42	1,12	39	1,24	33			6
7	1,41	28	1,24	36	1,12	40									7
	♏		[♊]		[♉]		[♈]		[♓]		[♒]		♑		

(Restorations have been made by comparison with
[23], f.164v, and consideration of tabular differences.)

بعد از جدول اختلاف ساعات بهر یک
از ساعات اختلاف و اختلاف عرض را
بگیرند و بیاساعات اختلاف بعد را بر ساعات
اجتماع افزایند اگر اجتماع غربی بود و از نقصان
اجتماع بکاهند اگر شرقی باشد یا ساعت وسط
کسوف حاصل شود بعد از آن عرض قمر را
در وسط کسوف حاصل کنند اگر شمالی تفاضل
میان او و میان اختلاف و اگر جنوبی باشد
جمع کنند یا عرض مرئی حاصل شود و اقل از نصف
و سه دقیقه باشد منکسف شود و الا فلا و چون
منکسف بشود بعض مرئی و علامت کسوف
عمل کنند جنانچه بعرض قمر و علامت خسوف
کرده بودند یا ساعات بدو کسوف وسط

f.34r

1   distance from the table of the difference of
    hours (parallax), each one
2   of the hours of difference and the difference
    of latitude,
3   obtain them. And the hours of difference of
    distance, add (them) to the time of the conjunc-
    tion
4   if the conjunction is westerly, and subtract
    them from the time of the
5   conjunction if it is easterly, so that the
    time of the middle of the solar
6   eclipse result. After that, the lunar latitude
7   at the middle of the solar eclipse is obtained;
    if (it is) northerly (take) the difference
8   between it and between the difference, and if
    it is southerly they are
9   added, so that the apparent latitude result.
    (If it is) less than thirty-
10 three minutes a solar eclipse will occur, and
    otherwise not. And when it is to be
11 eclipsed, do with the apparent latitude and
    the mark of the solar eclipse
12 as had been done with the lunar latitude and
    the lunar eclipse mark
13 so that the times of the beginning of the solar
    eclipse and the middle

ووسط کسوف والجلا معلوم شود و اصابع منخسفه
هیچ اصابع منخسفه بدانند و معرفت ساعات
اجتماع و استقبال در خاتمه با دانشا ان شاء الله تعالی
باب سیزدهم در معرفت وسط تحویل
از قبل تقویم شمس در وقت معین و ساعات
بعد تحویل مری عضاده را بر مثل تقویم شمس نهند
در وقت مفروض از اجزا دجره و حرف مسطره
را ابر کرشتعار کذ رانده موازی عضاده سازند
در بوضع تقاطع حرف مسطره بمحط صفیحه
علاماتی کنند پس حرف عضاده را برین علامت
نهند موضع مری عضاده از اجزا جره و وسط
تحویل بود پس استخراج وسط کنند درنصف
النهار مقدم بر وقت مفروض واین وسطارا

f.34v

1 and the middle[*] of the solar eclipse and
   clearance become known. And the solar
   eclipse digits

2 are determined like the lunar eclipse digits.
   And the determination of the hours of

3 conjunction and opposition, let it come in
   the conclusion, if God will, He is exalted!

4 CHAPTER THIRTEEN. On the Determination of the
   Mean of Transfer

5 by Means of the True Longitude of the Sun at
   a Given Time, and the Hours

6 after the Transfer. The pointer of the alidade
   is put according to the true longitude of the
   sun

7 at the assumed time on the divisions of the
   ring, and the edge of the ruler having been
   put

8 along the fictitious center, make it parallel
   to the alidade.

9 And at the place of intersection of the edge
   of the ruler with the circumference of the
   plate let a

10 mark be made. Then the edge of the alidade
    is put along this mark;

11 the place of the pointer of the alidade in
    divisions of the ring will be the mean of

12 transfer. Then the mean is extracted on the
    approaching

13 noon at the assumed time, and this mean,

---

* روسط, is repeated from the previous page.

از وسط تحویل کاهنداباقی رابرحرکت یکساعته

وسط قسمت کنند خارج قسمت بعد تحویل انصف

النهار مقدم باشد باب چهار دسم

درمعرفت ارتفاع حقیقی ازمری ومعرفت

ارتفاع مرئی ازحقیقی واختلاف سطر

بدایره ارتفاع مری عضاده رابرابر اول سرطان نهند

واز مرکز بجانب اول سرطان بجهت قمریک جز

ودو دقیقه بگیرند ازاجزاء عضاده وبجهت

شمس اگرآلت بزرگ باشد مشترک از دو دقیقه

بگیرند وبرای ازسره دو دقیقه بگیرند وعدد المتقی

علامتے کنند برصفحه سوا این را علامت سطر

خوانیم پس برجرف عضاده برمثل بعد شمس

وبا قمراز مرکز عالم دررمثل نصف بعد زسره

f.35r

1   let it be subtracted from the mean of transfer;
     the remainder is divided by the motion (in)
     one mean hour.

2   The quotient will be the distance of the trans-
     fer from the coming

3   noon. CHAPTER FOURTEEN.

4   On the Determination of the True Altitude from
     the Apparent and the Determination of the

5   Apparent Altitude from the True, and the [Paral-
     lax]*

6   in the Altitude Circle. The pointer of the
     alidade is placed at the first of Cancer,

7   and from the center, on the side of the first
     of Cancer in the case of the moon, one division

8   and two minutes is taken of the divisions of
     the alidade, and in the case of the

9   sun, if the instrument is large, a little more
     than two minutes

10   is taken. And for Venus two minutes is taken,
     and near the end

11   a mark is made on the plate, and we call this
     the mark of [parallax]*.

12   Then, on the edge of the alidade, according to
     the distance of the sun

13   or of the moon from the center of the universe,
     and according to half the distance of Venus

---

*MS. has علما ruler, for منظ .

از هر کوکب عالم علامتی کنند و ما این را علامت
کوکب نامیم پس اگر ارتفاع مری معلوم بوده باشد
و خواهند که ارتفاع حقیقی بدانند مری عضاده را
از اول حل بر توالی بقدر ارتفاع مری بگردانند
و حرف مسطره را بعلامت مسطر که داند بوازی
عضاده گردانند و بر صفیحه بقرب علامت کوکب
بر حرف مسطره خطی کشند پس عضاده را بگردانند
تا علامت کوکب بدین خط واقع شد و بعد مری
عضاده از اول حل ارتفاع حقیقی آن کوکب
باشد و تفاضل بین الارتفاعین اختلاف منظر
آن بود و بدایره ارتفاع و اگر ارتفاع حقیقی
معلوم باشد و ارتفاع مری خواهند مری عضاده
را بر توالی از اول حل بقدر ارتفاع حقیقی گردانند

f.35v

1   from the center of the universe a mark is
     made, and we call this the (parallax)

2   mark of the planet. Then, if the apparent
     altitude is known

3   and it is desired to ascertain the true longi-
     tude, the pointer of the alidade is turned

4   from the first of Aries along the succession
     (of the zodiacal signs) by the amount of the
     apparent altitude,

5   and, the edge of the ruler having been put
     along the parallax mark, turn it parallel to
     the

6   alidade, and on the plate at the planet's
     mark,

7   a line is drawn alongside the edge of the
     ruler. Then the alidade is turned

8   until the mark of the planet falls along this
     line, and the distance of the pointer of the

9   alidade from the first of Aries is the true
     altitude of [that]* planet.

10   And the difference between the two altitudes
     is its parallax

11   in the circle of altitude. And if the true
     altitude

12   is known and the apparent altitude is wanted,
     the pointer of the alidade is turned

13   in the direction (of the signs) from the first
     of Aries to the amount of the true altitude.

---

*For ا و read آن.

وبرموضع علامت کوکب که برعضاده کرده
بود نذ برصفیحه علامتے کنند پس حرف مسطر را
بابین علامت وعلامت مسطر کذرانند وعضاده را
موازی مسطر کردانند بعد مری عضاده ازاوّل
جل ارتفاع مری آن کوکب باشد والله اعلم

باب پانزدهم درمعرفت تعدیل الایّام
بلیلیها ایل این صناعت شبانه روزی حقیقے
رااز نصف النهار کیرند تا نصف النهار تا
باختلاف بقاع مختلف نشود پس مقدار
شبانه روز حقیقے دوری باشد ارمعدّل النهار
یا مطالع الخافیاب دران شبانه روز حرکت
کند خط استواو حرکت آفتاب مختلف است
چه کاهی بطئی السیرست وکاهی سریع السیر

f.36r

1 and at the place of the mark of the planet,
   which had been made on the alidade,

2 a mark is made on the plate. Then the edge of
   the ruler is placed

3 along this mark and the mark of parallax, and
   the alidade is turned

4 parallel to the ruler. The distance of the
   pointer of the alidade from the first of

5 Aries is the apparent altitude of that planet,
   but God knows better.

6 CHAPTER FIFTEEN. On the Determination of the
   Equation of

7 Time. Those having to do with this art take
   the true day (nychthemeron)

8 from noon to noon so that

9 in the difference of locality there will be
   no differing. So let the amount of the

10 true day be a rotation of the celestial equator (together)

11 with the rising (time) of that (arc through)
   which the sun in that day moves with respect
   to the (celestial)

12 equator. And the motion of the sun is different (i.e., variable),

13 since sometimes it is slow-travelling and
   sometimes fast-travelling,

و سر مطالع اجزاء فلک البروج مساوی نیست
پس مقادیر ایام حقیقی مختلف باشد نسبت
این بعد و تفاوت وا هل حساب چنانچه بایند
مشتا دی الاقدار جهت معرفت ع کا تا وط
کواکب غران پس آن زیاده بر دور را مساوی
سیر یکروز و سط آفتاب کرفته نا ایام سال
شکل فی ما بکشند و این ایام متساوی را ایام
وسطی خوانند و مقادیر آن دوری باشد از معدل
النهار ما قوسی مساوی سیر یکروز و سط آفتاب
و تفاوت میان این ایام و ایام حقیقی را تعدل
الایام بلیا لیها خوانند پس بجهت معرفت تعدل
الایام بلیا لیها تقویم شمس و سطش را حاصل
کنند در وقت مفروض پس بر وسط شمس جح

f.36v

1   and also the rising (times) of the divisions
     of the zodiac are not equal.

2   So the amounts of the true days are different
     with respect to

3   both differences, and those concerned with
     computation need days

4   equal in amount for the knowledge of mean
     motions of

5   planets, and for (problems) other than that.
     So that excess over a revolution has been
     taken equal to the

6   travel in one day of the mean sun, so that
     the days of the year be

7   sufficient, and these equal days are called

8   mean days. And the amounts of (each of) these
     is a rotation of the celestial

9   equator (together) with an arc equal to the
     travel of the mean sun in one day,

10   and the difference between these days and the
     true days is called the equation of

11   time. So, with regard to the determination of
     the equation of

12   time, the true longitude of the sun and its
     mean is obtained

13   at the assumed time. Then to the sun's mean
     three degrees

وبحار وسعت دقیقه واسے نایه زیادت
کنند و فضل مجموع ان را به مطالع تقویم شمس بگیرند
پس برای هر درجه ازین فضل چهار دقیقه از ساعتی
ساعات و برای هر ده دقیقه از دقایق فضل
یک دقیقه از ساعات و برای هر دقیقه از دقایق
فضل چهار ثانیه بکمیرند مجموع د قایس و ثوانی ساعات
تعدیل الایام بیابد لها بود از ایام و ساعات حقیقی
نقصان کنند ایام و سطی ماند والله اعلم
خاتمه در عمل بلوج اتصالات هر یک از بیب
معدل وبعد ماضی وساعات نصف النهار وساعات
لیل را بدانند پس از مسطر اطول بقدر ساعات
نصف النهار افزا کنند و راس مسطر لیل را که
در حزر نماست محاوی ساعات نصف لها را

f.37r

1 and fifty-seven minutes and thirty seconds is added,

2 and the excess of the sum of this over the (right) ascension of the sun is obtained.

3 Then, for each degree (of arc) of this excess four minutes of the

4 minutes of hours are taken, and for each ten minutes of the minutes of the excess (take)

5 one minute of hours, and for each minute of the minutes of the

6 excess four seconds are taken. The sum of the minutes and seconds of the hours will be the

7 equation of time. From the true days and hours

8 decrease them, the mean days will remain, but God knows better.

9 CONCLUSION. On the Operation of the Plate of Conjunctions. Each one: the daily rate,

10 and the past distance, and the time of noon, and the duration of

11 night, should be ascertained. Then extend the longer ruler to the amount of the time of

12 noon, and the head of the night ruler, which is

13 in the second trough, is put opposite the duration of the day

کنند از اجزاء خاسته لوح تا بعد راس سطر یوم
از راس سطر لیل بقدر ساعات نهار کرده و و
سطر غدات را بر شش ساعات لیل نهند از سطر
لیل پس زا بنه قایمه محاذی ساعات نصف نها
یوم آیند کرده از سطر غدات و آنچه از سطر
یوم محاذی سطر لیل است حکم محاسبه باشد بعد از ان
حرف سطر بدید بخط و عمل را مبدل از
اجزاء سرات بکند را خد و انگشت اسر قلم را بر شش بعد
ماضی از اجزاء سرات نهند و بر حطی که از ان جبر
خارج شده است بر انند یا کحرف سطر یا محیط
پس بخطی که از بنما با جزا آن ساعات و اجزا آ سطرهای
سه که فر و دی رود فرود آیند و نظر کنند که ان
خط جزار فطعه که در جکم محاست بر کدام فقیه

f.37v

1   on the divisions of the margin of the plate so
that the distance of the head of the day ruler

2   from the head of the night ruler is made (equal
to) the amount of the hours of the day. And

3   the next-day ruler is put according to the
hours of the night on the

4   night ruler. Then the right angle will be
opposite the hours of noon

5   of the coming day on the next-day ruler. And
that which of the

6   day ruler is opposite the night ruler let it
be disregarded. After that, the

7   edge of the turning ruler or thread is placed
according to the daily motion (<u>or</u> rate) in
the

8   divisions of travel, and the finger or pen-
point is placed according to the

9   past distance of the divisions of travel, and
is run along the line

10   extending from that division until (it reaches)
the edge of the ruler or thread.

11   Then descend along the line which runs from
here into the divisions of the hours and the
divisions of the

12   three-fold rulers. And note that that

13   line, except for the segment which is (to be)
disregarded, falls along which minute

اوکدام ساعات از کدام مسطره واقع است
آن ساعت اتصال از اول یوم یا اول لیل یا
از یوم آینده و موقع آن خط از اجزاء حاشیه
ساعات بعد است از نصف النهار مقدم
واگر مبدأ ساعات اتصال از یوم یا لیل
یا نصف النهار معلوم باشد و
بعد مجهول بعکس
ابن علی بعد را
بدانند
والله اعلم
م

TRANSLATION

f.38r

1   of which hour and which ruler.

2   This is the hour of conjunction from the
first (hour) of the day or the first (hour)
of the night

3   or of the coming day. And the place of that
line on the divisions of the edge

4   is the hours of the distance from the previous
noon.

5   And if the beginning of the hours of the conjunction from the day or the night

6   or noon is known, and

7   the distance unknown, by the reverse

8   of this operation the distance

9   may be ascertained,

10   but God knows better.

11   Finished.

COMMENTARY

References to the text and translation give the folio and line numbers of the Persian manuscript. Numbers enclosed in square brackets are references to the bibliography which begins on page 251.

1. The Introduction of the Manuscript (f. 2v:1 - 4v:4)

Folios 2v-4v are prefatory to the main body of the work, and the style and arrangement are typical of introductions to medieval Islamic scientific books. In contrast to the bald and unadorned language of the text proper these first passages are written in rhymed prose, in a florid and elaborate style. As was customary, the author chooses his figures of speech from the subject matter of his treatise, here astronomy.

The opening lines (f. 2v:1-10) are an invocation to God embellished with poetic allusions to the celestial portions of His creation. Assisted by a quotation from the Qur'ān, the author then modulates into a short (f. 2v:11 - 3r:2) panegyric on the Prophet.

Having thus paid his respects to established religion the theme is next announced - to enable the reader to solve astronomical problems without elaborate computation. The inventor on whom the anonymous author depends for his solution is named with fulsome praise, together with his two instruments.

The last section of the introduction is (f. 3v:13 - 4v:4) a dedication to the patron, the Ottoman Sultan Bayazid II (1447-1512), and a closing plea for forgiveness should the reader detect mistakes.

There follows (f. 4v:6 - 5v:7) a table of contents, after which the exposition proper begins.

2. Construction of the Disk and Ring (f. 5v:9 - 7r:1)

These two members together make up a large, 360° protractor, to be used for laying off all the angular distances needed in the subsequent operation of the instrument. The distinguishing peculiarity of this protractor, aside from its size, is the fact that its interior, the plate, is to be so constructed that it

can be rotated inside the ring, and fixed in any desired position. The manner of achieving this is shown in Figure 1,
(opposite f.11r) in which only a few details are conjectural.
The perforated tongue and the matching holes in the ring
prescribed in the Persian text are not mentioned in the Nuzhah.
Instead Kāshī prescribes (NS, p.270) that, the plate having
been set in proper position with respect to the ring, it be
fixed with a bit of wax. This detail is the only one in which
the author of the Persian text has exhibited any originality
whatever. In all other instances he simply chooses from among
the possibilities suggested by Kāshī. The use of a circular
set of fine, equally spaced holes is found also in Chaucer's
equatorium, but not for reproducing the apsidal motion. (Cf.
[42], pp.49-52.)

On our Figure 1 only four of the five circles demanded
by the text have actually been drawn. A circular scale showing
all five may be seen on page 48 of [42].

In these passages the language of the Nuzhah and that of
the Persian version are parallel, but with some divergence.
In speaking of the size of the instrument the former says
(NS, p.251)

> The least that it is possible for its
> diameter to be is half of the great cubit
> (al-dhirā⁣ᶜ al-kabīr), but it is better if
> it be two cubits of the Hāshimī cubits, or
> three cubits.

A cubit is the length of a forearm. The variety known as
the common cubit contains about 54.0 cm., the great Hāshimī
cubit about 66.5 cm., and the small Hāshimī cubit about 60.1 cm.
(Cf. [13], p.55; [49], p.491), thus the ambiguity of the texts
remains largely unresolved.

165

It is of interest to note that the same connection between
size of the instrument and precision made by our author (f. 5v:13
6r:2) is made also by Chaucer ([42], p.18), and in almost the
same terms.

3. The Jummal System of Writing Numbers

As was customary with Islamic astronomical works, the numbers in our text are represented by means of letters of the
Arabic alphabet. Each letter has a numerical value found by
coupling the letters, in the order of the old Semitic alphabet,
with the sequence 1,2,3,...,10,20,30,...,100,200,300,... .
Integers are indicated by proper combinations of letters in
what is thus a non-place-value decimal system.

Fractions, however, are displayed in a place-value system
with base sixty, the sexagesimal digits 0,1,2,...,59 being written
as indicated above. We transcribe them as ordinary numerals,
using commas to separate sexagesimal digits, and a semicolon
as sexagesimal point. When arcs are involved, units of thirty
degrees each may also be encountered, each one a zodiacal sign
(burj, plural buruj). For these we use a superscript s. Thus
either 9^s 23;7,0,54° or 293;7,0,54° means

$$293 + \frac{7}{60} + \frac{0}{60^2} + \frac{54}{60^3} \text{ degrees.}$$

Note that we write sexagesimals with the powers of sixty
in the denominators increasing to the right; in the Arabic script
they increase to the left.

To elevate (Persian rafᶜ kardan) a sexagesimal means, as
we would put it, to move the sexagesimal point one place to the
right. The first elevate (marfūᶜ) of 0;0,29,7. Its second
elevate is 29;7. To depress (munhat gereftan) a sexagesimal

is to perform the inverse operation of elevating, i.e., to
move the sexagesimal point to the left.

In a pure sexagesimal system, which we will use when
convenient, both integer and fractional parts of real numbers
are expressed in powers of sixty. For instance 1,0 = 60.

The reader will find additional material on the Arabic
sexagesimal system in [15].

4. Planetary Apogees (f. 7r)

For the moon and the planets, deferent circles are to
be drawn or engraved on the disk. The edge of the disk itself
is to serve as the sun's deferent. Kāshī, like most of the
other Islamic astronomers, regarded the planetary apogees as
fixed with respect to each other and to the fixed stars. It
was therefore necessary to impose the motion of precession on
all the apsidal longitudes. This was to be accomplished by
rotating the disk within the ring and setting it so that the
apsidal lines had the proper longitudes for the time for which
planetary positions were being determined.

The first step in the procedure was to assume a line of
apsides for the sun and then to lay off the other apsidal lines
with respect to it at the angular distances given in the table
on f.7r. The entries of this table are rounded-off values
from an equivalent table on f.128 of Kāshī's own Khāqānī Zīj
[23]. Our Table 1 below has in its second column these numbers
from the zīj and in the first column the longitudes of the
apogees according to this work as of the year 781 Yazdigerd,
the epoch of the zīj.

The numbers shown are completely secure, since positions
of the apogees are also given in the zīj for years 782, 783,...,
790 Yazdigerd, and scribal errors can be restored by checking

Planet	Apogee	Apogee of Each Planet Less the Solar Apogee
♄	8^S 11;27,26°	5^S 10;28,17°
♃	6 0;15,0	2 29;15,51
♂	4 17;4,26	1 16;5,17
♀	2 20;14,28	11 19;15,19
☿	7 3;39,28	4 2;40,19
☉	3 0;59,9	-----------

Table 1

one value against another.

From various sources, mostly Islamic, the editor has assembled a collection of thirty sets of planetary apogees. While many do not vary widely from this one, no other set is identical with it. It is probably taken from the Zīj-i Īlkhānī (No.6 in [29]) and may be based on observations made at the Marāghah observatory of Nasīr al-Dīn al-Tūsī.

5. Laying out the Planetary Deferents (f. 7r:10 - 8r:1)

After having located the apsidal lines, the next step is to mark the deferent centers. Each such center is on its respective apsidal line, at a distance from the plate center equal to the planet's eccentricity. A table of eccentricities is the first of the two tables on f. 7v, the units being sixtieths of the plate radius. The centers, having been marked permanently on the plate, the deferents can then be drawn, corresponding radii, expressed in the same units, being given in the second table on the same page.

It was customary to express planetary parameters in sixtieths of the deferent radius. Kāshī, however, gives them

all in sixtieths of the plate radius. He has shortened all
the deferent radii in order to prevent the circles from
running off the plate and to have them snugly nested, no two
intersecting. In order to "norm" these parameters, that is,
express them in the customary fashion, we have computed for
each planet a "norming coefficient", 1,0/R, where R is the
deferent radius as given in the text. The quotients make up
Column 2 of Table 2 below. They are used to norm the eccentricities of the text. Side by side with the results
(Column 4) are the eccentricities used in the Almagest. The
reader will note that corresponding parameters are identical,
or nearly so, except for Venus. The reading 1;3 is confirmed
in the Nuzhah and in the Khāqānī Zīj [23]. In the former all

Planet	① Deferent Radii	② = 1,0/① Norming Coefficients	③ Eccentricities	④ = ②·③ Normed Eccentricities	⑤ Ptolemaic Eccentricities	⑥ Epicycle Radii	⑦ = ②·⑥ Normed Epicycle Radii	⑧ Ptolemaic Epicycle Radii
♄	52;2	1;9,11	2;58	3;25	3;25	5;38	6;30	6;30
♃	55;28	1;4,55	2;32	2;44	2;45	10;38	11;30	11;30
♂	45;27	1;19,12	4;33	6;0	6;0	30;32	40;18	39;30
♀	58;58	1;1,3	1;2	1;3	1;15	42;25	43;10	43;10
☿	51;23	1;13,56	4;52	6;0	6;0	18;13	22;27	22;30

Table 2

the planetary parameters placed on the instrument are attributed

to the Īlkhānī observations. A non-Ptolemaic eccentricity for
Venus is not surprising. Nearby values used by al-Bīrūnī,
al-Zarqālla, and Ibn al-Shāṭir are cited in [31].

6. The Deferents of the Two Luminaries (f. 7r:4 - 8r:1)

There is no need to norm Kāshī's parameters as given for
either the sun or the moon. The solar deferent being the rim
of the disk, its radius is already sixty. As for the moon,
Ptolemy in the Almagest broke the 60-unit radius of the
Hipparchian non-eccentric deferent into two parts, of lengths
10;19 and 49;41. These are the lunar parameters given on
f. 7v.

Since the lunar apsidal line moves with great rapidity,
the disk must be specially set each time the moon's position
is to be determined, and the center of its deferent can be
marked at any convenient place on the disk, say along the
radius passing through the small tongue on its periphery.

The solar eccentricity prescribed, 2;6,9, differs
considerably from the Ptolemaic 2;30 ([43], iii, 4). There
is no doubt, however, as to the correctness of this reading.
It appears also in the Nuzhah (NS, p.252), and twice in the
Khāqānī Zīj ([23], ff. 95v and 157r). Doubtless this also is
from the Īlkhānī observations; in his zīj Kāshī simply says
that it is based on new observations.

It is close to the values used by other Islamic astronomers. Bīrūnī, in discussing the solar equation (cf. [30])
regards 2;5 as a rounded-off standard parameter. It is common
to Habash al-Hāsib, Abū al-Wafā', and the Banī Mūsā. Al-Battānī
([3], vol.i, p.213) made it 2;4,45. Ibn al-Shāṭir uses, in
effect, 2;7 (= 4;37-2;30, cf. [46], p.429).

7. The Deferent of Mercury (f. 8r:4-12)

The highly eccentric behavior of Mercury forced Ptolemy

to devise a model differing from that of the other planets. Instead of being fixed, the deferent center itself (D on Figure 3) moves on the arc of a circle with center at F and radius equal to three sixtieths of the deferent radius, DP. The equant center is at E, colinear with FC, C being the center of the universe, and CE = EF = FD. DP so moves that at all times $\Theta_1 = \Theta_2$. The resultant curve traced out by P and indicated by the succession of small circles on the figure is slightly oval, and has AR as an axis of symmetry. It can be regarded as a non-circular deferent for Mercury.

Kāshī approximates this curve by two circular arcs whose centers are at B and B' and whose radii are 51;23 sixtieths of the plate radius. As described in the text, B and B' are each 5;8 of these units on either side of the "turning center" (F). The latter, the center of the oval deferent, is to be 4;52 units along the line of apsides from the plate center.

In our Figure 3 the scale of the Ptolemaic locus has been so chosen that apogee A and perigee R coincide with the corresponding points on Kāshī's deferent. It will be observed that the latter is a good approximation to the Ptolemaic curve, the two loci practically coinciding for wide distances on either side of H and H'.

To obtain a norming coefficient for Mercury, we note that in Kāshī's units 56;0, the sum of the eccentricity (4;52) and the semimajor axis of the deferent (51;8), corresponds to 1,9;0 in Ptolemy's configuration. Hence the ratio of these two numbers, 1,9/56 = 1;13,56, is the desired coefficient.

8. An Oval Deferent for the Moon

In Appendix 1 of NS (p.289) Kāshī remarks that an oval

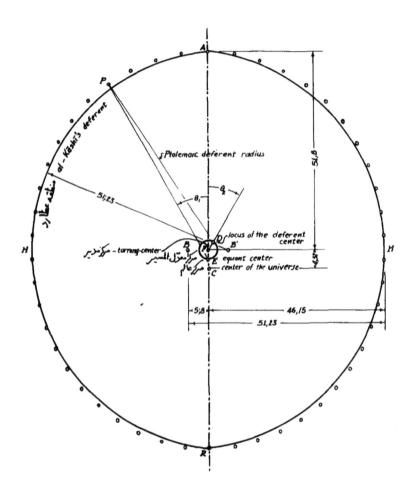

Figure 3. The Oval Deferent of Mercury

lunar deferent can be drawn permanently on the plate, thus
eliminating some of the manipulations, described in Chapter
II,4 of the text, for the determination of the moon's true
longitude. That this is the case can easily be verified by
recalling that in an abstract sense the Ptolemaic models for
the moon and Mercury differ only slightly (cf. [41], Appendix I).
The similarity is obscured by the rapid motion of the lunar
line of apsides, but with respect to the latter the epicycle
center does indeed trace out an invariant oval path in space.
It is curious that whereas a number of equatorium makers laid
out oval deferents for Mercury, insofar as we know only Kāshī
suggested doing likewise for the moon, and then only as an
afterthought.

The strongly (and erroneously) oval character of the
Ptolemaic lunar orbit is portrayed in Figure 4, which shows
the path of the moon plotted to scale at four day intervals
through the course of a month. The direction of the mean sun
on corresponding days is also indicated.

9. Marking the Equant Centers (f. 8v:1-7)

For each of the three superior planets and for Venus a
point is marked on the apsidal line, outward from the deferent
center by a distance equal to the eccentricity. For Mercury
the equant center (E on Figure 3) is put halfway between the
turning center (F), and the plate center (C). This corresponds
to the Ptolemaic arrangement.

The moon has no equant, but its "opposite point" is marked
on the plate, along the apsidal line, but on the opposite side
of the plate center from the deferent center. The distance
from the plate center to the opposite point is 10;19, the
lunar eccentricity.

173

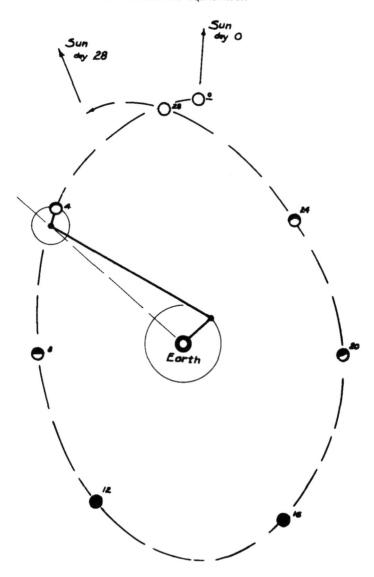

Figure 4.   The Ptolemaic Lunar Orbit

10. Alternative Plate Layouts

In the Nuzhah, Kāshī does not content himself with the
single arrangement of the plate and alidade given above. He
describes in addition variants of the basic scheme. For
instance, instead of letting the plate center represent the
center of the universe for all the planets it is possible to
make the deferent centers all coincide with the center of the
plate. Then for each planet except Mercury the points representing the center of the universe and the equant center are
placed symmetrically on the line of apsides on either side
of the plate center. A plate thus laid out is said to be of
the "parallel deferent" (mutawāzī al-manāṭiq) type.

Again, if no objection is made to the deferents intersecting, and if it is wished to retain the plate center as
center of the universe for all planets, full advantage may be
taken of the size of the plate to choose scales such that all
deferents will just touch the edge of the plate. This is
equivalent to demanding that the sum of the eccentricity and
the deferent radius equal 1,0. Tables of the resulting parameters are given in the Nuzhah, the units being sixtieths of
the plate radius.

Thirdly, in the "single deferent" (muttahid al-manāṭiq)
type the edge of the plate itself is made to serve as the
deferent for all the planets whose deferents are circular,
just as was the case with the sun. For each planet this
again pushes the center of the universe out along its individual
apsidal line, the eccentricities and epicycle radii being the
normed entries in Columns 4 and 7 of Table 2 above.

Finally, the plate may be laid out with the same apogee
for all the planets, and the position of the apsidal line
taken into consideration by a separate setting of the plate

within the ring for each planet. Or, since the apsidal motions
are slow, the movable ring may be dispensed with altogether,
the apsidal lines being laid out once and for all to a fixed
position in the zodiac correct for the time when the instrument is made.

Kāshī enumerates some fifteen combinations of these various
possibilities.

11. The Medieval Sine Function

The two references to sines in the text make it necessary
to state that the function referred to is not the modern function, for which the radius of the defining circle is unity,
but its ancestor, defined in terms of a circle of radius sixty.
We distinguish between the two by writing the symbol for the
medieval function with an initial capital, thus Sin θ, say.
The identity relating the two functions is

$$\text{Sin } \theta = 1,0 \sin \theta.$$

When computations are carried out in sexagesimals, the
transformation from one function to the other is trivial, since
the medieval function is simply the first elevate (cf. Section
3 above) of the modern function.

In our text the plate radius is consistently taken as
1,0.

12. The Latitude Circles on the Plate (f. 8v:10 - 10r:2)

The description in f. 9r:10 - 10r:2 is not found in NK at
all. It makes up part of Appendix 3 (p.291) of NS, hence
represents one of Kāshī's many afterthoughts, replacing in
the Persian text his original technique for solving the problem
of planetary latitudes. This shows that the anonymous author

176

COMMENTARY

of our text worked from a copy of NS.

The two versions are practically identical in meaning.
A permanent line, the equating diameter (shown on Figure 13)
is to be drawn on the plate and through its center. When
latitudes are being computed the plate is to be fixed in the
ring so that one extremity of the equating diameter coincides
with the origin on the ring, i.e. Aries 0°. Several semicircles are drawn permanently on the plate, all concentric
with the ring, and all having some segment of the equating
diameter as bounding diameter. The radius of the largest is
to be Sin 9° (cf. Section 11 above). A convenient way of
obtaining this is to join the pair of points on the ring which
are both nine degrees away from the equating diameter and on
the same side of it. The result is a line parallel to the
equating diameter and distant from it by Sin 9°. The required
semicircle can then easily be drawn, tangent to this line.
Then fill the inside of this semicircle with a set of parallel
lines as indicated in Figure 5 (reproduced from NS, p.292),

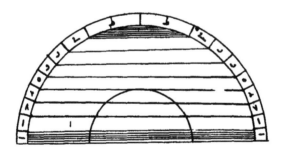

Figure 5. The Latitude Circles and Lines, from NS

these to be drawn by connecting pairs of graduations on the
ring which are at equal distances from the equating diameter.
The endpoints of the segments are to be marked with numbers
corresponding to the degrees distance of the point-pairs from
the equating diameter, 8°, 7°, 6°, and so on, and the lines at
fractional distances also. These segments are known as the
latitude lines. They are also shown, more crudely, on f. 11r.
The basic fact to be borne in mind concerning them is that if
a latitude line is marked Θ its distance from the equating
diameter is Sin Θ.

The second semicircle, the moon's latitude circle, is
to be drawn tangent to the 5° line. On Figure 5 (i.e. in the
NS) it has been made erroneously tangent to the 4° line.

The last two semicircles, for Venus' first latitude and
the latitude of Mercury, are drawn tangent to the ten minute
and forty-five minute lines respectively.

From a practical point of view, and on a plate of any
reasonable size, the Venus and Mercury semicircles would be
so small as to make it impossible either to draw or to utilize
them. Doubtless Kāshī was carried away by the theoretical
considerations described in Section 25 below.

For material on f. 10r:2-11 see Section 33 below.

13. The Latitude Points (f. 8v:12 - 9r:9)

Eight permanent marks are to be put on the equating
diameter, two for each superior planet, one each for Venus
and Mercury, all on the half of the diameter opposite the
first point of Aries. Distances of these points from the
plate center are given in the table on f. 9r. Following is
an explanation of how these particular distances were arrived
at.

COMMENTARY

In the planetary theory of the Almagest the deferent planes of all the planets are slightly inclined with respect to the ecliptic, intersecting the latter along the nodal line, which passes through the center of the universe. The nodal line of each inferior planet is perpendicular to the line of (deferent) apsides of that planet. For all the other planets, however, the angles between these two lines differ. Kāshī's values are given on f. 22v. They are the same as those of the Almagest ([43], ed. of Halma, vol.ii, p.414) except for Mars, for which the Ptolemaic value is 90°. But here the divergence is apparent only, for Ptolemy computes the angle as 95½°, subsequently deciding that use of 90° will involve only a negligible error. In effect Kāshī does the same thing.

Assume a line drawn in the deferent plane through E in Figure 12, the center of the universe, perpendicular to the nodal line MN. and intersecting the deferent in F (on the side of the apogee) and G (on the side of the perigee). Then, in the case of the superior planets, the two distances of the latitude points are EF and EG.

These distances can be computed directly in terms of the deferent radius and eccentricity. In all probability, however, Kāshī obtained them directly from the Almagest. For Saturn the corresponding distances in the Ptolemaic configuration are 1,2;10 and 57;40. For Jupiter they are 1,2;30 and 57;30. Division of each of these numbers by the proper norming coefficient obtainable from Column 2 of Table 2 gives the corresponding entries in the table on f. 9r.

As for Mars, its line of apsides is so close to FG that for two-place accuracy it is sufficient simply to add and subtract Kāshī's value of Mars' eccentricity from its deferent radius to obtain the desired quantities. Thus

$$EF = EA = ED + DA = 50;0$$

and
$$GE = PE = PD - ED = 40;54.$$

For both inferior planets, maximum inclination of the
epicycle plane occurs at the nodes (M and N on Figure 12).
Hence the distances of the latitude points for these planets
are ME and NE. But, AP being perpendicular to MN, ME = NE.
And the eccentricity of Venus is so small that its deferent
radius, 58;58, is used for the common distance. For Mercury,
46;0 is the distance perpendicular to the line of apsides
from the center of the universe to the deferent, hence the
distance to its latitude point.

14. The Alidade and Ruler (f. 11v:1 - 12r:6, 13r:4-8)

Although most of Chapter I,4 of the text is clear, there
are a few obscure passages which are best clarified by first
reading corresponding sections in the Nuzhah. The following
is a free rendition of NS, pp.253-255, interspersed with such
comments as seem appropriate. Make two rulers, says Kāshī,
one like the alidade of an edged (muharrafah) astrolabe, but
having two graduated edges, one for the true longitudes, the
other for latitudes. It is as though a pair of graduated
alidades were combined, back to back. Each edge shall have
a semicircular projection at its midpoint, as small as practicable, and pierced by a round hole so that either edge can be
pivoted to rotate about the plate center. In length the
alidade should be slightly greater than the plate diameter.

The first of the two edges, called the diameter-edge
(harf al-qutr), is to be graduated in equal divisions, sixtieths
of the plate diameter and fractions of these to the extent
possible. The divisions are to be numbered outward from the

center in both directions.

The second edge, the <u>arcs-edge</u> (<u>harf al-qussi</u>), has four
sets of graduations marked upon it, each set being the projections on the alidade edge of the points of division of the
ring. The four scales are obtained by effecting the projection
when the alidade is in each of four different positions. For
all four positions the alidade is pivoted on the plate at the
hole on the arcs-edge side. One extremity of the arcs-edge
is designated the head, or northern end, the other the tail
or southern end. If the arc from the head of the arcs-edge to
the first point of Aries on the ring is Θ, then the point on
the ring having a longitude of λ will project onto the edge at
a distance $\cos (\lambda + \Theta)$, provided that the plate radius is taken
as unity. The four prescribed positions can be fixed by spe-
cifying the value of Θ corresponding to each. They are

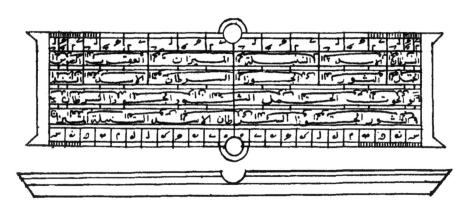

Figure 6. The Alidade and Ruler, from NS

1. $\theta = 90°$, for computing the latitudes of the
    moon and the second latitude of the inferior
    planets
2. $\theta = 0°$, for the latitude of Mars and the first
    and third latitudes of the inferior planets
3. $\theta = 80°$, for the latitude of Saturn
4. $\theta = 10°$, for the latitude of Jupiter.

The general appearance of the alidade is well portrayed in
Figure 6, reproduced from NS, p.256. Except for the double
pivot feature, it closely resembles the alidades on many
existent astrolabes, for example those shown in [9], pp.121,
162, 267, and 308.

The basic principles employed in the arcs-edge, that of
a linear nomogram for trigonometric functions, appears also in
the dastūr quadrant (cf. [50], p.73).

The second object described in Chapter I,4, the ruler
proper, is in NS specified as being narrower than the alidade,
but of the same length. As indicated in Figure 6, it is to
have a semicircular recess let into the center of one edge, of
the same diameter as the projecting lugs on the alidades, so
that the two straight-edges can be fitted snugly together.
The ruler edge is to be graduated in sixtieths of the plate
radius.

All three sources prescribe connecting the ruler and
alidade with a chain, and the mounting of sights on the alidade
to enable the user to take altitudes with the instrument. The
only other equatorium into which this observational feature
has been incorporated is the "Albion" developed in 1326 by
Richard of Wallingford. (Cf. [42], p.128).

Descriptions of the other markings on the alidade and

182

ruler are deferred to the sections where the solutions of
specific problems are discussed.

We are now in a position to return to the Persian text,
bearing in mind that its anonymous author has here abridged
the corresponding parts of the Nuzhah, sometimes so drastically
as to be unintelligible. He evidently has abandoned the
second set of graduations, the arcs-edge, hence he has no
need of the second semicircular lug on the other side of the
alidade.

As will be seen in the sequel, the main function of the
ruler is to mark out directions parallel to a setting of the
alidade edge made by using the graduations on the ring. According to f. 19v:9-11, the two edges are to be made parallel by
seeing that the arcs of the ring intercepted between them are
equal. Kāshī evidently noted the practical inconvenience of
this operation, and in Appendix 4 of NS (p.297) he describes
the construction of a set of parallel rulers. Our Figure 7
is reproduced from page 298 of NS.

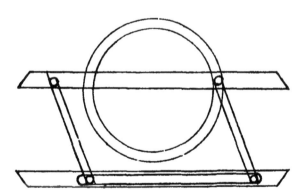

Figure 7. Parallel Rulers, from NS

15. The Difference Marks - Epicycle Radii (f. 12r:7 - 12v:3)

A set of six marks is to be engraved on one edge of the alidade at distances from the plate center specified in the table on f. 12r. Alternatively six circles may be drawn permanently on the plate, concentric with it, and having radii whose lengths are the tabular entries. In either event, each number is the length of the epicycle radius of the planet with which it is associated.

For the moon, the radius of 5;17 is close to the Ptolemaic 5;15. It is rounded off from Kāshī's own 5;16,46,36 resulting from the eclipse observations described in the opening sections of his zīj ([23] f. 4r - 6r).

Since the deferent radii of the planets proper have been shortened below the standard 60 units, their epicycle radii also have been cut down in proportion. In order to compare them with the lengths used by other astronomers it has been necessary to multiply each one by the corresponding norming coefficient found in Column 2 of Table 2. The results appear in Column 7 of the same table. It will be noted that only Mars and Mercury exhibit any difference with the Almagest values. In his zīj ([23], f. 98v) Kāshī states that the length 40;18 was determined as a consequence of the Ilkhānī observations. For Mercury the Ptolemaic 23;30 reappears on f. 110r of his zīj.

For material on f. 12v:3 - 13r:3 see Sections 36 and 40 below.

16. The Table of Mean Positions and Mean Motions (f. 13r:9 - 13v:6, 14v, 18r:9 - 19v:3)

The raw data to be fed into the instrument consist of the mean positions of the planets and their anomalies at a given

instant. These numbers are to be computed from some such table
as that supplied on f. 14v of the text. The column headings
of this table are self-explanatory except for the last two.
The term khāseh-yi murakkabeh, compound anomaly, seems to be
peculiar to Kāshī. He means by it the sum of the mean solar
longitude and the mean anomaly of an inferior planet.

The entries in the first ten rows of the table give values
of the mean longitude or anomaly for the initial instants of
years 851, 852, 853,..., 860 Yazdigerd respectively. According
to the marginal note on f. 18r the first day of 851 Yazdigerd
(= 16 November, 1481) was chosen for epoch as having been the
year of the enthronement of Bāyazīd II. The epoch of the era
of Yazdigerd is 16 June, 632 A.D., the years being of just
365 days, made up of the twelve thirty-day months named in
the table, plus five epagomenal days. (Cf. [10], p.298).
The entries in the remaining rows display the motion of the
respective longitudes or anomalies in the lengths of time
indicated. These are 10, 20, 30,..., 100, 200, 300,..., 1000
Yazdigerd years, then months, days, and one hour. The numbers shown for each month are cumulative in the sense that
they represent the total motion from the beginning of a year
up to the beginning of the particular month. The entries
opposite the numbers 1, 2, 3,..., 10 are cumulative in the
same manner, for example, those in the same row as 4 give
the motion for three days. The entries in the row of the
second 10, however, are motions for ten days, and those in
the following row are for twenty days.

All entries are in zodiacal signs, degrees and minutes,
and integer multiples of twelve signs have been dropped from
the reckoning. Numbers in square brackets indicate restorations to the text. The technique of checking entries is

explained in Sections 17 and 18 below. The only systematic
errors in the table are those entailed by the computer having
taken an erroneous 12;3,38° per year (instead of the correct
12;13,38°) for the annual mean motion of Saturn. This has
affected only that part of our table which is not identical
with the mean motion table of the Nuzhah.

The process of determining mean positions for a given
time is explained in Chapter II,2 of the text, and requires
no amplification except perhaps for the correction to be
applied when a determination is to be made for a place of
longitude other than that of Constantinople. Under such
circumstances the difference in degrees between the longitudes
of the two places is multiplied by 24/360 = 0;4 to obtain the
time difference. This is added to the local time if the
locality is east of Constantinople, otherwise it is subtracted.

The longitude of the latter place as given in the text,
60°, is probably rounded off from the value of 59;50° given
in the Khāqānī Zīj ([23], f. 74v). Al-Battānī ([3], vol.ii,
p.44) has 56;40° for the same coordinate, following Ptolemy
(cf. [14], p.196). Al-Khwārizmī, however, gives 49;50°,
which appears in numerous places in Ms. Vat. Gr. 1058. Hence
it looks as though Kāshī's zīj value is a miscopied version
of this, which our text rounds off. This still leaves us
with two distinct traditions for the longitude of Constantinople.

17. "Squeezing" Mean Motion Parameters from Tables

Although the entries of our table are given to three sexagesimal places only, it is possible to "squeeze" from the
tabular information a good approximation to the much more
precise elements used to compute the table in the first place.
The technique can best be explained by a worked example. We

ngCOMMENTARY

deduce below the base parameter for Mercury's compound anomaly.

A first approximation to the daily motion is seen from
the table to be 4;6° per day. Multiplication by 60 = 1,0
moves the sexagesimal point one place to the right to yield
4,6;0° for the distance travelled in two Persian months. The
entry opposite Khurdād, however, is 8ˢ 5;33°, so we correct
our first approximation to 4;5,33° for the daily motion. The
product of this by 365 = 6,5 gives 24,53;45,45°. Casting off
complete revolutions of 360° = 6,0°, there remains 53;45,45 ≈
1ˢ 23;46°. The most frequent tabular difference between
entries in the yearly positions is 1ˢ 23;44°, so we correct
our yearly distance to 24,53;44° and multiply by 10. The
result is 4,8,57;20°, the terminal digit of which is corrected
to 19 by comparison with the tabular entry of 5ˢ 27;19°, and
the process continues. It ends with a value of 6,54,55,31;52°
for the distance travelled in 1000 Persian (i.e. Egyptian)
years. Now operate backwards, successively dividing by 1000,
then by 365 to obtain 24,53;43,54,43,12° and 4;5,32,41,42,27°
for the basic yearly and daily rates of this parameter.

Thus the process is seen to be one of gradually building
up the higher digits of the parameter by repeated multiplication, each time correcting the lower digits by comparison
with the table. Each parameter has been treated in this
manner to obtain the set exhibited below.

18. The Mean Motion Base Parameters

As squeezed from the table and expressed in degrees per
day the mean motions used in the text are

☉		0;59,8,19,44,36°
☽		13;10,35,1,53
☽	(anomaly)	13;3,53,56,30,54

187

♌		0;3,10,38,18,44
♄		0;2,0,36,4,44
♃		0;4,59,16,23
♂		0;31,26,39,35,54
♀	(anomaly)	0;36,59,28,13
☿	(anomaly)	3;6,24,21,58
Apogees (=precession)		0;0,0,8,27

The anomalistic motions shown for the two inferior planets do
not result directly from the table. They are obtained by
subtracting the solar motion from the compound anomaly.

Comparison of these numbers with the set of parameters
independently squeezed from the Khāqānī Zīj ([23], ff. 127v -
130r) by M. Agha enables us to verify the author's statement
that the positions and motions are indeed those of the zīj.
The tables of the zīj, however, do not give mean longitudes
as such, but the mean "centers", that is, distances of the
mean planets from their respective apogees. The epoch of the
zīj, moreover, is 781 Yazdigerd, and its base location is
Shīrāz. The longitude of the latter place is taken as 88;0°
(cf. [23], f. 73r).

We note further that in general these are the parameters
of the Īlkhānī Zīj, upon which Kāshī's tables are based. For
the moon, however, Kāshī has computed independent parameters
based upon his observations of three lunar eclipses, those of
2 June and 26 November, 1406, and 22 May of the following
year. ([23], f. 4r - 6r).

19. Mean Motions in the Nuzhah

Both versions of the Nuzhah refer to a table of mean
motions. It is missing from the only extant copy of NK, but

appears on p.259 of NS. The epoch of this table is different
from that of our text; it gives mean positions for Yazdigerd
years 801, 802, 803,..., 810, and the base longitude is 88°,
that of Shīrāz (cf. Section 18 above). Otherwise the two
tables are the same.

In Appendix 5 of NS Kāshī states that he has worked out
a device for eliminating even the few additions and subtrac-
tions involved in computing mean positions from a table. His
invention is in essence a circular slide rule for performing
algebraic addition of arcs. To construct it he prescribes
drawing a family of circles on one face of the plate, concentric with it, and in sufficient number to accomodate the
scales next described. These amount to a graphical mean
motion table. For all planets the mean positions given in
the table are marked on the circles, the origin of each circle
being its intersection with the plate radius drawn to the
first point of Aries on the divisions of the ring. In addition to this, arcs are laid out on suitable circles, their
lengths being equal to the mean travel of the various planets
and their anomalies in hundreds, and tens of years, in months,
days, and fractions thereof to the extent to which marking
the plate is practicable. A metal ring is made so as to fit
snugly on the plate just touching the inner edge of the
graduated ring surrounding the plate. Two straight rulers,
of length a little greater than the plate radius, are pivoted
at the plate center in such fashion that an edge of each
passes through the plate center. If the date for which a
planet's mean position is required happens to be the beginning
of one of the years marked on a circle, simply project this
mark with the edge of one of the rulers out to the graduations
of the outer ring. Otherwise use the two rulers and the mean

motion scales to lay out on the inner ring an arc equal to the
mean travel of the planet in question in the time elapsed
between the given date and one of the years permanently marked.
Then rotate the inner ring so that the initial point of the
marked arc lies along the projection of the fixed point just
utilized. The desired mean position will now be at the
projection on the outer ring of the arc's terminal point.

Kāshī remarks that if each of the circular scales is made
so that it can be individually rotated, there will be no need
for the inner ring.

For material on f. 15r:1 - 17v see Section 43.

20. Determination of the Solar True Longitude (f. 19v:4 - 20r:4)

The Ptolemaic model of the sun's motion being extremely
simple, the determination of its true position with the instrument is equally simple. The plate circumference itself represents
the solar deferent, the apogee (represented in Figure 1 and 8
by the plate tongue) having already been fixed at its proper
longitude. The fictitious center (F in either of Figures 1 or
8) stands for the center of the universe. Thus CF, the distance
from the plate center to the fictitious center, is the solar
eccentricity. Having computed the sun's mean longitude, lay
it off as the arc AM by using the divisions of the ring. M
is Kāshī's mark of the mean. Then the angle at which M is
observed from F, measured from the equinoctial direction, is
the true longitude. To evaluate this angle, lay the edge of
the ruler along FM and rotate the alidade until it is parallel
to the ruler. Then the arc AT is the required true longitude.
The angular difference between the true and the mean longitudes,
arc MT, is the solar equation. The length of MF, measured

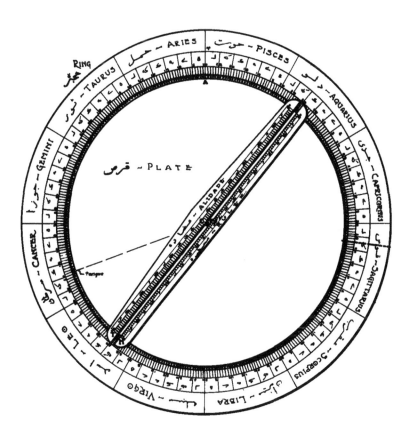

Figure 8.   The Instrument, Set Up for Finding
the Sun's True Longitude

with the scale marked off on the ruler, is the earth-sun
distance. In this case there is no need for a norming operation since the units are the standard ones in ancient astronomy,

191

sixtieths of the deferent radius.

21. Determination of the Lunar True Longitude (f. 20r:5 - 20v:13)

In using the instrument to find the true ecliptic position
of the moon at a given time, first compute from the mean motion
table: (1) the lunar mean longitude, (2) the lunar mean anomaly,
and (3) the solar mean longitude. Then mark P (see Figure 1),
the position of the mean moon, on the edge of the ring. Find
e, the moon's mean elongation, being the difference between
(1) and (3) above. Starting from P measure clockwise along
the edge of the ring an arc PS equal to 2e, the double elongation. Rotate the plate until its tongue comes opposite S and
fix it here with the peg. By means of the alidade mark E the
intersection of CP with the moon's deferent. E is Kāshī's mark
of the center. Put the ruler along E and N, the point opposite,
and rotate the alidade until it is parallel to the ruler. Mark
B, the intersection of the alidade edge with the edge of the
ring, the directed segments CB and NE thus being parallel and
in the same sense. As the author says, it is from this point
that the mean anomalistic motion is to be measured. Now rotate
the alidade clockwise the amount of the anomaly, (2) above,
through the arc BK. This puts the alidade in the position
shown in our figure.

In order to complete the Ptolemaic lunar configuration
all that remains is to lay off the epicycle radius in proper
length and direction (parallel to the alidade edge) from E.
Its endpoint L will then correspond to the position of the
moon in space. However, this construction presents some
difficulty from the practical point of view, for, as is the
case with our figure, L may run off the plate entirely. Moreover, we are not primarily interested in locating L, but rather

in determining the direction of the vector CL. We recall
that a lunar difference mark (Cf. f. 12r:9) has been put on
the alidade edge at a distance from the center equal to the
epicycle radius. The author specifically cautions us that
the position of the alidade should now be such that the
difference mark falls opposite the end of anomalistic motion,
i.e. D being the difference mark, vector CD must have a sense
opposite to vector EL. Mark the position of D on the plate
and lay the ruler along DE. Since the latter equals the
required vector sum of CE and EL, if the alidade is now
rotated parallel to the ruler, its intersection, G, with
the edge of the ring will give the required true longitude
of the moon. The final passage in the chapter (f. 20v:12-13)
applies only if the plate has not been set in the ring to show
the proper lunar apsidal longitude at the start of the operation.

22. Determination of a Planetary True Longitude (f. 21r:1 - 21v:7)

The solution of this problem with the equatorium resembles
that of the analogous lunar problem. First compute from the
table of mean motions the mean longitude of the sun for the
instant required, and either the planet's mean longitude (for
a superior planet) or the compound anomaly (for an inferior
planet). Presumably the plate has already been fixed inside
the ring so that the solar and planetary apogees have their
proper longitudes. Letter references in the sequel are to
Figure 9, which shows the final positions of alidade and ruler
in the solution of such a problem for the planet Mars. The
drawing is to scale.

Rotate the alidade until its edge is at the planet's mean

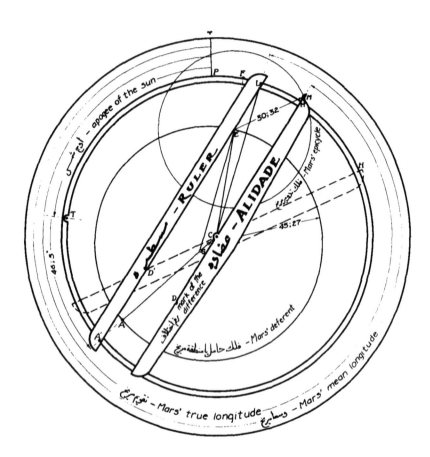

Figure 9. The Instrument, Set Up for Finding
Mars' True Longitude

longitude, L, on the divisions of the ring. Place the ruler
so that its edge is alongside the planet's equant-center, G,

and simultaneously parallel to the alidade edge. Where the ruler intersects the deferent, E, make the __center-mark__. This locates the planet's epicycle center at the given time.

For a superior planet turn the head of the alidade, that end of it opposite the side on which the difference marks (cf. f. 12r:9) have been made, until it reaches the mean solar longitude, H, and mark on the plate the point where the difference mark, D', then lies. Now put the ruler so that its edge lies along D and E, and rotate the alidade until it is parallel to the ruler. Then the intersection M of the head of the alidade with the divisions of the ring gives the required true longitude. For the vector D'C has been constructed in magnitude and direction equal to the vector EM from the epicycle center to the planet, and side CM of parallelogram CD'EM gives the direction of the vector sum of CE and D'C. And since EM is parallel to CH, EM has the required direction of the mean sun.

For the inferior planets, Venus and Mercury, the construction is of the same sort, bearing in mind that their mean longitude is the mean longitude of the sun. The same figure may be used to illustrate the configuration, although of course it will no longer be to scale. Now L is the sun's mean longitude and arc PA'H is the compound anomaly.

23. The Planetary Equations (f. 21v:8 - 22r:11)

The term __equation__ (ta'dīl) was in general used in ancient and medieval astronomy to denote a correction, in general small, applied to a function representing some phenomenon. The usage has survived in modern astronomy in such expressions as "the equation of time".

A planet's equation in longitude was defined as the difference between its true and mean longitudes. The equation was

195

in turn made up of two components. The <u>first equation</u>, or <u>equation of the center</u>, is that caused by the eccentric equant; the <u>second equation</u> is the effect of the planet's motion on the epicycle. (Cf. [4], p.96.) Clearly the algebraic sum of the first and second equations equals the equation of the planet.

All <u>zījes</u> (astronomical handbooks) contained tables of both equations, essential for the computation of true longitudes. In addition, since the second equation is dependent on the first, a fairly involved interpolation scheme was necessary, to estimate the effect of the first equation on the second. It is the great advantage of an analog computer such as the equatorium that true longitudes are determined without the interposition of the equation or its components at all.

Nevertheless, should the user want to find the equations with the instrument he can do so by following the instructions in Chapter II,6. The author may have had in mind an individual who is computing a position accurately with a zīj, but who wants a quick, crude check of his partial results.

For whatever reason, having made the prescribed marks, at L, the mean position in Figure 9, and F, the projection of the epicycle center on the ring, it is clear that the arcs LF and FM are the first and second equations respectively.

When a mean position was measured from the apsidal line of the particular planet it was called the <u>center</u>, here arc A'HL. Addition of the first equation to the center gave the adjusted center, here arc A'HF.

The closing statement of the chapter, that if the adjusted mean longitude is subtracted from the sun's mean (for the superior planets), or from the compound anomaly (for the inferior planets), the adjusted anomaly remains, is equivalent to the valid expression

$$PA'H = PA'F + FA'H.$$

24. The Lunar Equations (f. 22r:2)

In the ancient and medieval lunar theory the terms first equation and second equation did not denote angles analogous to those associated with the same terms in the planetary theory.

The lunar anomaly was not laid off from the epicyclic apogee of the model. Ptolemy found it necessary instead to count the anomaly from an epicyclic apogee as determined with respect to the "point opposite" (N on Figure 1), not with respect to the center of the universe, C. (Cf. [41], p.195). The first lunar equation was defined as the angular displacement in the epicyclic apogee caused by this situation. (See [23], f. 78v). On Figure 1 it is angle CEN. And since CH has been constructed parallel to EN, angle BCP also equals the first equation. But this angle, being a central angle on the plate, is measured by arc PB. P is evidently the "second mark" of our passage in the text, and B is the "mark of the beginning of the anomalistic motion", i.e. the statement in the manuscript is valid.

The second lunar equation was defined as the effect of the anomaly on the mean longitude, provided that the epicycle center were at maximum distance from the center of the universe. Since, in general, the actual distance was less than the maximum, in using the lunar tables it was necessary to add to the second equation a suitable correction. On the configuration on the instrument, however, this second equation does not appear as such, and the author of the manuscript makes no mention of it.

25. The Latitude of the Moon (f. 22r:13 - 22v:7)

The moon's orbit can be thought of as lying in a plane which makes a fixed angle of about five degrees with the plane of the ecliptic. The line of nodes, the intersection between the two planes, is not fixed, but rotates slowly in a negative direction with respect to the vernal point. Let β be the moon's latitude, λ its longitude, and λ_n the longitude of the ascending node, the ecliptic point through which the moon passes in going from southern to northern latitudes. Then we have

$$(1) \qquad \sin \beta = \sin 5° \sin (\lambda-\lambda_n) = \sin 5° \sin \omega,$$

an exact spherical-trigonometric relation involving ω, the _argument of the latitude_, the moon's ecliptic distance from the ascending node.

In computing lunar latitudes Ptolemy did not use (1), but the equivalent of the expression

$$(2) \qquad\qquad\qquad \beta = 5° \sin \omega,$$

a reasonably good approximation. (Cf. [43], ed. of Halma, vol.i, p.316).

We have already described how to find λ and λ_n. Once put on the ring of the instrument, their differences can be noted directly. All versions of the text (f. 22r:13, NS p.277) say add, rather than subtract, λ and λ_n. This usage is explained by recalling that the motion of the node is contrary to the order of the zodiacal signs. Hence the nodal positions are to be plotted, as we would say, negatively, and algebraic subtraction in this case becomes addition.

Having determined ω, to complete the operation rotate the

198

alidade until its pointer crosses the graduations of the ring
at the point corresponding to ω. Then note at which of the
latitude lines the edge of the alidade crosses the moon's
latitude circle described in Section 12 above. The result is
the lunar latitude.

The operation is pictured in Figure 10, where we note that

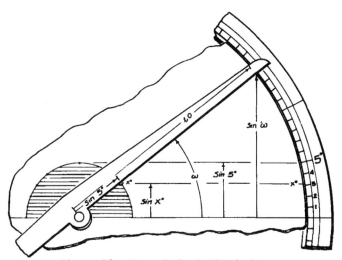

Figure 10. Use of the Latitude Lines

by similar triangles and use of the properties of the latitude
lines (cf. Section 12),

$$\frac{Sin\ x}{Sin\ \omega} = \frac{Sin\ 5^o}{1,0} \ ,$$

or

(3)   sin x = sin 5° sin ω.

Now comparison of expressions (1) and (3) shows that x = β,

199

that is, Kāshī's process utilizes the exact expression for
determining the lunar latitude. In theory it is better than
the result of using expression (2).

In the original version of the Nuzhah (NS, p.277) the
method of determining the lunar latitude is as follows.
Elevate the alidade so that it makes an angle of five
degrees with the equating diameter. Mark on the edge of
the alidade the point corresponding to ω on the lunar latitude
scale. Project this point parallel to the equating diameter
on to the divisions of the ring. The number on this scale
corresponding to the projection is the lunar latitude. It
is easy to show that this process also yields the result of
applying (3).

The Chaucer equatorium manuscript [42] gives a method
of determining the lunar latitude as resulting from (2).
There is no provision for the computation of planetary latitudes.

26. Planetary Latitudes in the Almagest

Chapter II,7 is by far the largest section of our text,
taking up about a quarter of the entire second treatise. That
this is the case is not surprising, for the topic of which it
treats, planetary latitudes, is the complicated result of a
complicated phenomenon. In determining planetary longitudes
it is convenient to regard all motion as taking place in the
plane of the ecliptic, which has the effect of making the
problem a two-dimensional one. This cannot be done with the
motion in latitude. The accompanying theory assumes that
the planes of both deferent and epicycle make small but non-
zero angles with the ecliptic and with each other, some of
the angles being variable.

200

C O M M E N T A R Y

Ptolemy regarded the latitude of a planet at any instant
as being the algebraic sum of two or three component parts,
known as the <u>first latitude</u> (β_1), <u>second latitude</u> (β_2) and,
in the case of the inferior planets, the <u>third latitude</u> (β_3).
In order to define these components it will be necessary for
us first to introduce a number of other terms.

It was customary to designate as <u>first diameter</u> of the
epicycle, the line of true epicyclic apsides, that is, the
diameter joining the true epicyclic apogee and perigee. In
Figure 11, BC, B'C', and B"C" are three positions of the first
diameter. The <u>second diameter</u> of the epicycle is the diameter
perpendicular to the first ([4], pp.61, 64). Examples from
the same figure are DF, D'F' and D"F".

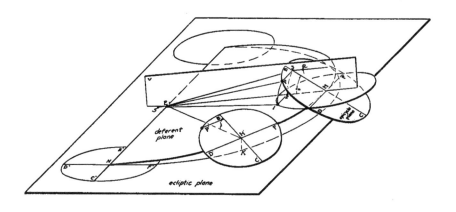

Figure 11. The Tilted Deferent and Epicycle

The first latitude is the angle made by the ecliptic plane

with the line joining the center of the universe and the epicycle center. In other words, β_1 is the portion of the total
latitude due to the deviation (hereafter denoted by i) the
angle at the line of nodes between the deferent and ecliptic
planes.

The second latitude is the component due to the inclination (denoted by j), the tipping of the epicycle about its
second diameter, the latter being maintained parallel to the
ecliptic plane in the case of the superior planets.

With the two inferior planets, however, a third latitude
is involved, caused by the obliquity, a tilting of the epicycle
about its first diameter.

A useful concept is that of the center of latitude, denoted
by $\bar{\omega}$ and defined as being the distance on the ecliptic from
the ascending node of a planet to the true longitude of its
epicycle center. In Figure 12 the expression

$$\bar{\omega} = \angle NEC = \angle NEA + \angle AEC$$

holds identically for all positions of C and A. Angle AEC is
called the adjusted center and is denoted by $\bar{\lambda}_a$. In the case
of the inferior planets, $\bar{\omega} = \bar{\lambda}_a + 90°$, the line of apsides and
the line of nodes being for them perpendicular. For the other
planets the angles between these two lines are given in the
table on f. 22v.

Chapter II,2 can be regarded as comprising four sections.
The first, f. 22r:13 - 22v:7, disposes of the lunar latitude.
This we have already commented on, in Section 25. The second
section, f. 22v:7 - 25r:11, the longest and most involved of
the four, describes simultaneously the determination of the
latitudes of the superior planets and β_2 for the inferior
planets. We separate the commentary on this section into two

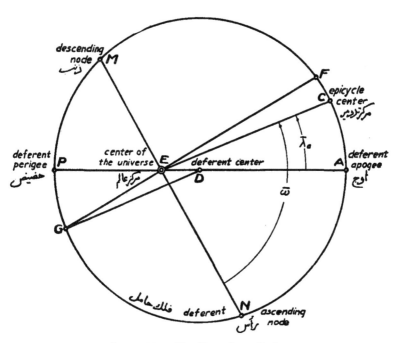

Figure 12. The Planetary Nodes

parts, Sections 28 and 30 below. The third section, f. 25r:11 -
26r:7, explains how to determine β_3 for an inferior planet and
is discussed by us in Section 29. Finally f. 26r:7 - 26v:6
has to do with β_1 for an inferior planet. As being the simplest
operation of the planetary latitude group, we describe it in
the section immediately following this.

27. The First Latitude of the Inferior Planets (f. 26r:7 - 26v:6)

The geometric model used in the Almagest for the inferior
planets has the deferent plane seesawing through a small angle
north and south of the ecliptic plane about an axis in the

ecliptic plane perpendicular to the (deferent) line of apsides and passing through the center of the universe. The deviation of the deferent plane is given by the expression (cf. [8], p.199)

(4) $i = i_m \sin \bar{\omega}$, $i_m = \begin{cases} 0;10° \text{ for Venus,} \\ -0;45 \text{ for Mercury.} \end{cases}$

Then, if in Figure 11, H' is the location of the epicycle center at a given time, in the spherical triangle NH'K, arc H'K is β_1, the desired first latitude, angle H'KN is a right angle, and since i (angle H'NK) is small the approximate equality

(5) $\beta_1 \approx i \sin \bar{\omega} = (i_m \sin \bar{\omega}) \sin \bar{\omega} = i_m \sin^2 \bar{\omega}$

subsists. This is the Ptolemaic theory.

The construction described in our text yields a result which is very slightly different. The alidade is elevated from the equating diameter by an angle equal to $\bar{\omega}$. We then note at which of the latitude lines the edge of the alidade inter-sects the proper latitude circle (cf. Section 12). Suppose the designation of this latitude line is x. This number will then satisfy the expression

(6) $\sin x = \sin i_m \sin \bar{\omega}$,

the transformation being of the same sort as that described for the lunar latitude in Section 25 above.

Now put the alidade perpendicular to the equating diameter and mark on its edge the point where the x latitude line crosses it. The distance from the mark to the plate center will then be Sin x. Return the alidade to its original position so that it makes an angle $\bar{\omega}$ with the equating diameter, and note the latitude line on which the mark falls. If this is the y lati-tude line y is the desired first latitude.

The second part of the operation is an iteration of the first with i_m replaced by x and x replaced by y. Hence we have

(7) $\sin y = \sin x \cdot \sin \bar{\omega} = (\sin i_m \cdot \sin \bar{\omega}) \sin \bar{\omega}$

$\qquad = \sin i_m \cdot \sin^2 \bar{\omega}.$

Comparison of expression (5) with (7) and the small size of i_m verifies the approximate equality of the results obtained with the two expressions.

In practice the construction is completely out of the question because of the minute semicircle required to be drawn on the plate. Even from a strictly theoretical point of view, what was justified in the case of the lunar latitude cannot be defended here. In both cases a right spherical triangle was involved, the solution of which involved an expression analogous to (6). But (4) is a device for giving a simple harmonic motion, and is not an approximate equality obtained from an accurate expression such as (6).

28. The Second Latitude of the Inferior Planets (f. 22v:7 - 25r:11)

In Figure 11 three positions of a planetary epicycle are shown, its center lying at N, H, and H'. In the first position the inclination j, the angle the first diameter makes with the radius vector from the center of the universe to the epicycle center, is zero. In the second case it has taken the maximum value, j_m, and the third case illustrates a general position intermediate between the two. In all three situations the second diameter (D"F", DF, and D'F') is shown parallel to the ecliptic plane, hence β_3 and the obliquity are zero. Values

of j_m for each planet are to be found in the table on f. 23r.
They are identical with those of the Almagest ([43], ed. of
Halma vol.ii, p.255). In the transcription we have put a minus
sign in parentheses before the entry for Venus so that the ulti-
mate result will have the proper sign, north being taken as
positive. A precise and general statement of the value of j,
angle B'H'E, for all positions on the deferent, is

(8) $j(\bar{\lambda}_a) = j_m \sin \bar{\lambda}_a = j_m \sin (\bar{\omega} - 90°) = - j_m \cos \bar{\omega}$

for the inferior planets, and

(9) $j(\bar{\omega}) = j_m \sin \bar{\omega}$

for the superior planets. Note that for an inferior planet
maximum inclination occurs when the epicycle center is on the
nodal line, in contrast to the situation pictured in Figure 11.
This figure, however, shows all the elements affecting the
latitude $(\beta = \beta_1 + \beta_2)$ of a superior planet. For the intermediate position shown, $\bar{\omega}$ is the arc NK, the inclination, angle
B'H'E, is

$$(\angle BHE) \sin \angle NEK,$$

and β is the angle P'E makes with the ecliptic plane. Also,
if the relative position of the ecliptic plane there shown is
ignored, the same figure may be used to illustrate the determination of the second latitude of an inferior planet.

   Ptolemy regarded as prohibitively complicated a direct,
general computation and tabulation of β_2 as a function of two
variables, $\bar{\lambda}_a$ and α. He therefore adopted the simplification
sketched below, the ensuing sacrifice of accuracy not being
large. He computed β_2 for general values of α, but for $\bar{\lambda}_a$ = 90°,

i.e. for the position of the epicycle center at H which gives maximum inclination of the first diameter, hence maximum β_2. Here this max $\beta_2(\alpha)$ is the angle PE makes with the deferent plane. He then defines the second latitude in general as

$$(10) \qquad \beta_2(\bar{\lambda}_a, \alpha) = \max \beta_2(\alpha) \cdot \sin \bar{\lambda}_a,$$

expressed in modern notation.

Kāshī finds max β_2 for a general α by means of a clever construction in the manner of descriptive geometry in which the single plane of the instrument's plate is regarded as containing the planes of the deferent, the plane denoted by v in Figure 11, and the plane of the epicycle. The equating diameter (UC in Figure 13) represents the intersection between v and the deferent plane, with H (the center of the instrument) representing the center of the epicycle. E is the latitude point (cf. Section 13) of the particular planet being dealt with, so placed that EH equals the distance (along the perpendicular to the line of nodes) from the center of the universe to the deferent of the planet. The plane v is folded down into the plane of the plate, about EH, in such fashion that its trace with the epicycle plane takes the position H3. And the plane of the epicycle is rotated into the deferent plane through an angle of j_m about its second diameter DF so that its first diameter takes the position BC.

If, now, the true length of PE (in Figure 11) can be constructed, as well as the perpendicular from P to the plane of the deferent, then the problem will be as well as solved. For if a right triangle is constructed with PE as hypotenuse and altitude equal to the perpendicular just referred to, the acute angle at its base will be the desired max β_2.

To this end the alidade is put so that its pointer crosses
the ring at the graduation corresponding to the amount of the
anomaly (α). The "first mark" (1 in Figure 13) is then made
on the plate at the position of the permanent difference mark
on the alidade edge. This makes H1 equal in length to the
epicycle radius of the planet. Now make the "second mark" (2
on Figure 13), being the projection of 1 on the equating diameter.

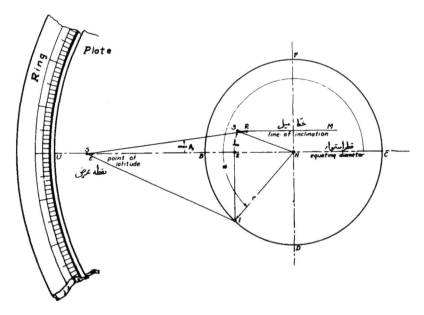

Figure 13. Construction for the Planetary Latitudes

By use of the alidade make the "third mark", 3, on the plate
so that angle 3H2 = j_m and H2 = H3. Through point 3 draw a
line M3 parallel to the equating diameter. The distance between
this pair of parallel lines is the altitude of the desired right
triangle.

208

To find the hypotenuse of the triangle mark point S the "substitute for the latitude point" on the equating diameter so that S2 = E3. It is as though, when the epicycle is rotated into the deferent plane, the right triangle EP3 also is flattened down into the deferent plane without change of size, the vertex E being made to slide outward along EH, 3 taking the position 2, and P the position 1. Then point S1 is the true length of EP in Figure 13.

Now, with center S and radius S1 draw an arc cutting M3 in R. Then the angle RSH is the desired max $\beta_2(\alpha)$, and it can be measured in the usual manner by placing the alidade parallel to SR and noting the point on the divisions of the ring where the pointer of the alidade falls.

The final steps in the determination consist of performing an operation analogous to the lunar latitude determination, the only differences being that the lunar latitude semicircle of radius Sin 5° is replaced by a mark on the alidade edge whose distance from the plate center is Sin max $\beta_2(\alpha)$. Thus the resulting β_2 from the instrument is given by the expression

$$(11) \qquad \sin \beta_2(\bar{\lambda}_a, \alpha) = \sin \max \beta_2(\alpha) \cdot \sin \bar{\lambda}_a,$$

which leads to a result differing slightly from that of (10). The criticism made of the β_1 determination in Section 27 above applies also to this section of the text.

The passage concludes (f. 25r:3-25r:11) with the usual complicated rules for determining the sign of the result. Negative numbers were not known to the medieval scientists. In cases where a result involved one of two directions, the result itself having issued from a combination of two or more elements, each with a direction of its own, there was nothing for it but to enumerate all possibilities in a set of rules.

· The term _parecliptic_, occurring in this passage, requires
explanation. Following Nallino ([3], vol.ii, p.352) we translate by it the word _mumaththal_, which is a contraction of
al-falak al-mumaththal li-falak al-burūj. It is a sort of
reference circle, one for each planet, lying in the plane of
the ecliptic and having its center at the center of the universe.
Apparently the Islamic astronomers made use of the concept
because they thought of the universe in terms of a set of
more or less concentric and overlapping shells, each shell
containing the deferent and epicycle of a planet. On the
parecliptic of each planet were projected the positions of
that planet. If in any context the word _ecliptic_ is substitu-
ted for _parecliptic_ the essential meaning will be unchanged.

29. The Third Latitude of the Inferior Planets (f. 25r:11 - 26r:7)

In approximating the final component of the latitude
Ptolemy neglects the simultaneous but small effects on it of
the first two components. Thus in Figure 14, which shows the
upper half (BPC) of the epicycle tilted about its first diameter (BC), the plane of EON' may be considered either as the
plane of the deferent or as that of the ecliptic. E is the
center of the universe, P a general position of the planet
corresponding to an anomaly of α. EM is tangent to the epicycle,
and the right spherical triangle OM'N' represents a portion of
the celestial sphere, N' being the projection of M' on the
ecliptic arc ON'. OM, which is approximately equal to q, its
projection on the ecliptic, is the "second equation" of the
planet. Clearly q is a function of α.

The obliquity defined in Section 26 above is measured by
the spherical angle M'ON'. This angle varies sinusoidally as

210

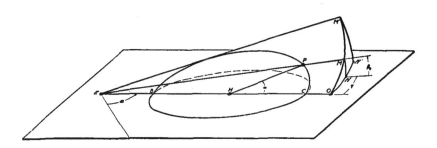

Figure 14.   The Epicycle Tilted for the
Third Latitude Component

H travels around the deferent, being zero when H is at the
nodes and a maximum or minimum when it is in the line of apsides.

Ptolemy treats the spherical triangles MON and M'ON' as
though they were plane triangles, using what is tantamount to
the crudely approximate relation

$$\frac{MN}{q} = \frac{M'N'}{ON'} \ ,$$

or

(12) $$\beta_3 = MN \approx q\left(\frac{M'N'}{ON'}\right) = qk$$

for any fixed obliquity. The number k depends on the size of
the epicycle, the deferent, the eccentricity and the maximum
obliquity. Making the additional assumption that, for fixed
q, β_3 varies directly as the obliquity, angle M'ON', his Almagest
tables ([43], ed. of Halma, vol.ii, p.414; [3], vol.ii, p.250)
are computed as though

211

$$(13) \quad \beta_3(a,\bar{\omega}) = kq \sin \bar{\omega}, \quad k = \begin{cases} 1/18;24 & \text{for Venus} \\ -0;6,8 & \text{when } 0° \leq \bar{\omega} \leq 180° \\ -0;7,30 & \text{when } 180° \leq \bar{\omega} \leq 360° \end{cases} \Bigg\} \begin{array}{l} \text{for} \\ \text{Mercur} \end{array}$$

were the general definition of the third latitude for Venus and
Mercury. The opposite signs for k imply opposite tilting of the
epicycle planes in the two cases. The two different values of
k for Mercury are to compensate roughly for the eccentric deferent
of this planet. As the epicycle leaves the ascending node it
travels toward the apogee, hence, in general it is then farther
from the center of the universe than after leaving the descending
node. Thus the same obliquity produces less effect at the greater
distance, so the value of k is smaller in absolute value on the
apogee side of the line of nodes. The eccentricity of Venus is
so small that the corresponding effect for it is neglected.

Except for differences in parameters, Kāshī's method is a
graphical duplication of the whole scheme. The text prescribes
taking a third of a sixth of Venus' second equation, i.e. 1/18.
For Mercury one is to multiply the second equation by either
0;7° or 0;8° depending respectively on whether the epicycle
center is or is not on the same part of the deferent as is the
apogee.

It is to be noticed that in two cases Kāshī's values for
k differ from the Ptolemaic ones by as much as a digit in the
first sexagesimal place.

It then remains only to impose on the extreme obliquity
thus found the now familiar sinusoidal transformation which
employs the latitude lines on the plate. The final result is
thus implicit in the expression

$$(14) \qquad \sin \beta_3(a,\bar{\omega}) = \sin kq \cdot \sin \bar{\omega},$$

which is very nearly equivalent to (13).

30. Latitude of the Superior Planets (f. 22v:7 - 25r:11)

The Almagest arrangements for the superior planets are
simpler on two counts than those for the inferior planets. For
one thing, there is no obliquity, hence $\beta_3 = 0$. And secondly,
both β_1 and β_2 vary (roughly) sinusoidally with $\bar{\omega}$, not one with
$\bar{\omega}$ and the other with $\bar{\lambda}_a$ as above. In Figure 11, for instance,
when the epicycle center is at N both β_1 and β_2 are zero; at H
both are maximal. In the general position H', β_1 equals the
great circle arc KH', which is approximately equal to $i_m \sin \bar{\omega}$.
Ptolemy therefore computes as previously max $\beta_2(\alpha)$ by finding
the angle EP makes with the deferent plane. Now he takes

$$\max \beta(\alpha) = \max \beta_1 + \max \beta_2(\alpha) = i_m + \max \beta_2(\alpha)$$

for general values of α. Finally he defines

(15) $$\beta(\alpha, \bar{\omega}) = \max \beta(\alpha) \cdot \sin \bar{\omega}.$$

But, in contrast to the inferior planets, the deferents of
the superior planets are not symmetrically placed on the line
of nodes; in general, if the epicycle center is on the northern
part of the deferent (i.e. $0° < \bar{\omega} < 180°$), it will be farther
from the center of the universe than when it occupies a corresponding place on the southern portion ($180° < \bar{\omega} < 360°$). Hence
both Ptolemy and Kāshī use two different values for EH (Figures
11 and 13) for each superior planet, depending on whether it is
in the first or last two quadrants. These pairs of distances
are the distances of the latitude points given in the table on
f. 9r and discussed by us in Section 13. Except for this,
Kāshī's construction for max $\beta_2(\alpha)$ is as explained in Section
28 above. To this add the corresponding constant i_m. Then
perform the customary transformation on it involving $\sin \bar{\omega}$ to
obtain the β implicit in the expression

(16) $\sin \beta(a, \bar{\omega}) = \sin \max \beta(a) \cdot \sin \bar{\omega},$

which differs very slightly from (15) for the small β involved.

One passage, f. 23v:3 - 23v:11, is puzzling. It looks as though the author wants the line of inclination to make an angle of i_m with the equating diameter. This seems pointless. It is true that i_m is involved, but it should be added to the max $\beta_2(a)$, angle RSH, not to the line of inclination. Also, he seems to want the line of inclination cut off in length equal to the epicycle radius. This would do no harm, but it appears unnecessary.

31. Planetary Distances (f. 26v:13 - 27v:12)

For any given time, the distance from the earth to the moon is the length of DE in Figure 1, where this figure represents the lunar configuration at the given instant. In like manner, for any planet the earth-planet distance is the length of line D'E in Figure 9. These distances having been measured in sixtieths of the plate radius, to convert them into distances measured in the standard scale, sixtieths of the respective deferent-radius, it is necessary to multiply each by a proper norming coefficient. These are given by the author in the table on f. 27v and by us in Table 2, Column 2. Determination of the solar distance has already been discussed, in Section 20 above.

32. Stations and Retrogradations (f. 28r:1 - 30r:1)

When a planet's forward motion in the zodiac ceases, it is said to be muqīm, stationary. It then becomes retrograde (rājiᶜ), having passed through the first, or retrograde station (maqām-i rajᶜat). After a time it again becomes stationary, passing through the second or direct station (maqām-i istiqāmat); thence it resumes forward motion and is said to be <u>direct</u> (<u>mustaqīm</u>).

Our author devotes a chapter to the use of the Plate of Heavens in computing the time of reaching a station. He follows implicitly the directions given in the Nuzha, which is based on the theory of Ptolemy, who in turn bases his development on a proposition he attributes to Apollonius of Perga (c. 200 B.C.). In substance, this elegant theorem ([43], ed. of Halma, vol.ii, p.312) states that if the deferent center and the center of the universe coincide, then the planet (P in Figure 15) will be stationary when

$$\frac{m}{n} = \frac{v_e}{v_p} \; ,$$

v_e being the angular velocity of the epicycle center E about O, v_p the angular velocity of P about E, and OT being perpendicular

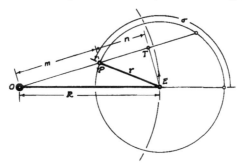

Figure 15. Diagram for Apollonius' Theorem

to ET. The <u>station</u> (<u>maqām</u>) is σ, the value of the epicyclic anomaly when the above expression obtains. Of course this is a simplification of the actual Ptolemaic model, for in the latter the center of the universe is displaced from the deferent center by a distance d. So in general OE is not a constant, but a function of λ, here the longitude of the epicycle center measured from the deferent apogee. We call this variable radius $\rho(\lambda)$, and note that $\rho(0°) = R + d$ and $\rho(180°) = R - d$. Moreover, Apollonius' theorem applies only approximately to this model, and the location of the station on the epicycle is also a function of λ, $\sigma(\lambda)$ say. Ptolemy computes directly only three values of this function for each planet: $\sigma(0°)$, $\sigma(180°)$, and $\sigma(v_d) = \sigma$, now the location of the station when the epicycle center is at the mean distance. For intermediate values of the argument he uses what amounts to the interpolation scheme

$$(17) \quad \sigma(\lambda) = \begin{cases} \sigma + \dfrac{[\rho(\lambda) - R][\sigma(0°) - \sigma]}{d}, & 0 \le \lambda \le v_d, \\[2ex] \sigma + \dfrac{[\rho(\lambda) - R][\sigma - \sigma(180°)]}{d}, & v_d \le \lambda \le 180°. \end{cases}$$

Thus he makes the change in $\sigma(\lambda)$ proportional to the change in $\rho(\lambda)$ for corresponding λ. (Cf. [3], vol.ii, p.246).

The arrangements in our manuscript are somewhat different, at least in appearance. On f. 29r is a table of $\sigma(0°)$ and $\sigma(180°)$, reproduced below together with the corresponding values determined by Ptolemy ([43], ed. of Halma, vol.ii, p.355), al-Battānī ([3], vol.ii, p.138), and Ulugh Beg [54]. It should be remembered that these numbers depend on four parameters, not only on d and r, but also on the mean motions and mean anomalistic rates of the planets involved. Kāshī makes no mention or use of an independently computed $\sigma(v_d)$; instead he puts

$$(18) \quad \sigma(\lambda) = \sigma(0°) + \frac{[\rho(0°)-\rho(\lambda)][\sigma(180°)-\sigma(0°)]}{2d} .$$

The difference between this and expression (17) is more apparent than real. In fact, if $\sigma(v_d) = [\sigma(180°)-\sigma(0°)]/2$, they are equivalent, and Ptolemy's values for $\sigma(v_d)$ are practically the mean between his extreme values.

	$\sigma(0°)$				$\sigma(180°)$			
	Almagest	Kāshī	Ulugh Beg	al-Battānī	Almagest	Kāshī	Ulugh Beg	al-Battānī
♄	$3^s22;45°$	$3^s22;45°$	$3^s22;10°$	$3^s22;45°$	$3^s25;29°$	$3^s25;29°$	$3^s24;49°$	$3^s25;29°$
♃	$4^s 4;5°$	$4^s 7;6°$	$4^s 3;40°$	$4^s 4;5°$	$4^s 7;11°$	$4^s10;11°$	$4^s 6;51°$	$4^s 7;11°$
♂	$5^s 7;28°$	$5^s 7;14°$	$5^s 5;36°$	$5^s 7;33°$	$5^s19;9°$	$5^s18;48°$	$5^s19;42°$	$5^s19;14°$
♀	$5^s15;51°$	$5^s15;45°$	$5^s16;12°$	$5^s15;53°$	$5^s18;21°$	$5^s18;27°$	$5^s18;4°$	$5^s18;21°$
☿	$4^s27;14°$	$4^s24;29°$	$4^s27;14°$	$4^s27;13°$	$4^s24;40°$	$4^s27;14°$	$4^s24;40°$	$4^s24;40°$

Table 3

The values laid out on the plate for $\rho(0°)$, $\rho(180°)$ and
2d are displayed on f. 28v for the convenience of the user. $\rho(\lambda)$,
the "preserved distance", is to be obtained by direct measurement with the ruler.

The first and second stations, being symmetrically disposed
on the epicycle with respect to the true (epicyclic) apogee,
computation of the one for a given λ gives the other immediately.

The final step in the determination consists simply of
evaluating $(\sigma(\lambda)-\alpha)/\dot{\alpha}$ where α is the anomaly for the time being
and $\dot{\alpha}$ is the rate of change of α with respect to time. The
result will be the time until station is reached, in the same
units used to express $\dot{\alpha}$.

Special remarks concerning Mercury (f. 28v:5, 29r:2) are
occasioned by the oval deferent of this planet (see Figure 3),
the effect of which is to bring the epicycle nearest the center
of the universe at two different points, not at the deferent
perigee as is the case with all other planets. These two points
are about 120° away from the deferent apogee, as is inferred

217

from Kāshī's rule, and as can be verified by inspection of the
last column in the Almagest table of Mercury's equations ([43],
ed. of Halma, vol.ii, p.309).

33. The Planetary Sectors (f. 10r:2-11, 10v, 30r:2 - 31r:10)

The subject of Chapter II,10 in our text is treated in
[28], which may be consulted for details by the reader. We
content ourselves here with a minimum for understanding of the
text, beginning with a restatement of the definitions with which
the chapter begins.

For certain astrological purposes, it was customary to
divide the deferent and the epicycle into four sectors each,
called by our author apogee sectors (nitāqāt-i aujī) and
epicyclic sectors (nitāqāt-i tadvīrī) respectively. There
were two categories of sectors, those computed according to
distance, and those according to velocity (harakah), or, as
our author puts it, movement (sayr).

In all cases the beginning of the first sector was at the
(deferent or epicyclic) apogee, the beginning of the third
sector at the perigee.

The beginnings of the second and fourth distance sectors
on either the deferent or the epicycle were defined as those
points at which an object moving on the circle in question is
at its mean distance from the center of the universe.

For the velocity sectors, the beginnings of the second
and fourth are those positions of a point moving on the deferent or epicycle at which its angular velocity, as viewed from
the center of the universe, attains its mean value.

The endpoint of each first sector coincides with the
beginning of the second, and so on.

The table on f. 10v locates the sector boundaries for all

218

categories. The entries have evidently been obtained from
Kāshī's zīj ([23], f. 141v), uniformly rounded off to minutes.
We next address ourselves to the question of how these values
were determined.

The problem is simplest for the sun, which has no epicycle, hence no epicyclic sectors, and no equant. If d and
R are the deferent eccentricity and radius respectively, then
the arc of the deferent from the beginning of the first sector
to the beginning of the second, that is, the amplitude of the
first solar distance sector is

$$(19) \qquad 90° + arc \sin \frac{d}{2R} .$$

This disposes of all other solar sectors of this category,
for, as always, the sectors are symmetrically located with
respect to the line of apsides. Hence the amplitudes of the
first and fourth sectors are always equal, while the first and
second are supplementary.

The amplitude of the first solar velocity sector is

$$(20) \qquad 90° + arc \sin \frac{d}{R} .$$

With the planets, the location of the initial point of the
second deferent distance sector can also be obtained from (19)
above. For use in the zījes, however, it was convenient to
tabulate the angle subtended by the first sector at the equant
rather than at the deferent center. For small d a good approximation to the former angle is

$$(21) \qquad 90° + arc \sin \frac{3d}{2R} .$$

For a planetary deferent velocity sector the angle subtended
by the first sector at the equant is approximately that given

by expression (20) above.

For the tabulated values of the epicycle distance and velocity sectors the expressions

(22)
$$90° + \text{arc sin } \frac{r}{2(R+d)} \text{ ,}$$

and

(23)
$$90° + \text{arc sin } \frac{r}{R+d}$$

respectively have evidently been used by the original computer of the zīj, where r is the epicycle radius.

In the zīj itself the "equation" also is tabulated, that is, the variation in the results when R+d is replaced by R-d. Something of the sort is necessary to take account of the variation in the distance from earth to epicycle center, a variation which causes small changes in the size of the sectors. For intermediate positions between the extremes of R+d and R-d an interpolation scheme was used which is not mentioned in our text. It is to this variation, however, which the author has reference in his statement on f. 30v:1.

Independent computations using Kāshī's parameters in the expressions above have resulted in a verification of all the entries in the text's table of sectors except for some involving Mercury and the moon, which, having special models, demand special treatment.

For Mercury's deferent distance sectors, the last column in the Almagest table of Mercury's equations ([43], ed. of Halma, vol.ii, p.309) measures the variation in the maximum size of the equation due to the epicycle as compared with its mean value. Hence any point at which this function vanishes marks a position at which the epicycle is at mean distance, i.e. the initial point of the second deferent distance sector.

Interpolation in this table yields a zero when the argument is
67;13°, a result reasonably close to our tabular value of 67;44°.

As for the deferent velocity sectors of Mercury, (19) above
is applied as though it were valid for a non-circular as well
as a circular deferent. Expressions (22) and (23) are used for
the epicycle sectors of both Mercury and the moon.

Kāshī's definition of the initial point of the second lunar
deferent distance sector is evidently taken to be the value of
the double elongation at which the epicycle center will be at
mean distance from the center of the universe. This is

$$\text{arc cos } \frac{d}{2R} = \text{arc cos } (\frac{10;19}{2(49;41)}) = 84;2°,$$

as required by the table.

In the Almagest table of lunar equations ([43], ed. of
Halma, vol.i, p.316) the value of the double elongation which
gives maximum displacement of the epicyclic apogee (Column 3 in
the Almagest table) is 114°. This is our tabular entry for the
moon's deferent velocity sector. In fact, however, the lunar
epicycle travels on the deferent in such fashion that its
angular velocity as viewed from the center of the universe is
a constant. Hence the concept of deferent velocity sectors as
defined above has no meaning. In some way Kāshī must have
associated the 114° with the conditions which identify the
mean angular velocity of the epicycle centers of the planets,
but how he did so is not clear.

So much for the table. As for the instrument, since all
deferents are marked on the plate, it is an easy matter to mark
also the boundaries of the deferent sectors, as prescribed in
the text (ff. 10r:2, 30v:2).

The epicycles as such do not appear on the instrument,

221

and to determine the epicycle sector of a planet at a given
time recourse must be had to manipulations with the instrument.
The author's directions in f. 30v:5 - 31r:2 have reference to
a person standing at the periphery of the horizontally placed
instrument, opposite the planet's epicycle center, and looking
toward the center of the instrument. If the position of the
planet is on his right this implies that the true longitude
exceeds the adjusted mean longitude (i.e. the longitude of the
epicycle center) and that the planet is in either the first or
second epicyclic sector. If it is on the left it is in the
third or fourth. Compare the distances of the planet and the
epicycle center from the center of the universe. If the first
exceeds the second the planet is in either the first or fourth
epicyclic distance sector. If the reverse is the case it is
in either the second or third. The combination of these two
criteria suffices to locate the planet's epicyclic distance
sector. The author gives no directions for the velocity
sectors.

The terms "increasing" (or "increased") and "decreasing"
(or "decreased") as used in the last part of this chapter
(f. 31r:3-9) are explained by Bīrūnī in [4](p.203). "Increasing in computation" means that the equation is positive. The
author is here referring to the epicycle sectors only. "Increased
in magnitude" refers to the celestial object's apparent size,
which varies inversely with its distance. When it is in the
second or third epicyclic sector its apparent size will be
increased over its mean value, hence the term.

34. Prediction of Lunar Eclipses (f. 31r:11 - 33r:3)

A lunar eclipse occurs whenever the moon enters the shadow
cast by the earth. If for any time during the eclipse the moon

is completely immersed in the shadow the eclipse is said to be
total, otherwise it is partial. Our author seeks to describe
eclipses to the extent of determining the times of (a) first
contact of moon and shadow, (b) first totality, if the eclipse
is total, (c) the middle of the eclipse, (d) the beginning of
clearance, i.e. the end of totality, and (e) complete clearance.
For a partial eclipse the magnitude also is desired.

Since the depth and duration of an eclipse is determined
by how closely the broken line sun-earth-moon approaches
straightness, it is clear that a necessary condition for a
lunar eclipse is that sun and moon be in opposition, i.e. have
longitudes differing by 180°. The condition is not sufficient,
however, for the moon does not travel on the ecliptic, but
along a second great circle which intersects the ecliptic at
an angle of about five degrees. This circle rotates slowly
westward so that the nodes, the points of intersection between
the two circles, have a motion of about nineteen degrees per
year. The moon's distance from the ecliptic, its latitude, is
the element which determines whether or not an eclipse will
occur at a given opposition.

Our author makes several simplifying assumptions. For
one thing he regards the earth-moon distance as a constant,
hence the apparent size of the lunar disk as viewed from the
earth is also constant. The same sort of assumption for the
earth-sun distance implies that the width of the shadow cone
where it is cut by the moon is likewise a constant. These
simplifications have the effect of making an essentially three-
dimensional problem two dimensional, for now we can confine
our attention to the surface of the sphere, concentric with
the earth, in which the moon's center moves. As a final
simplification, let Figure 16 represent to some scale a small

223

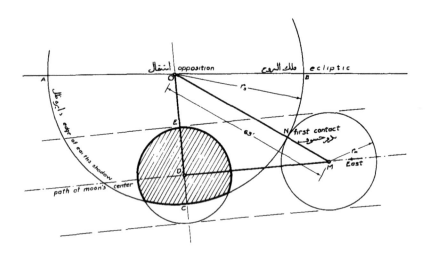

Figure 16. A Diagram for a Partial Lunar Eclipse

portion of this spherical surface, with O marking a point on
the ecliptic AB at which moon and sun are in opposition. The
path of the moon's center is MD and its apparent radius is MN.
The intersection of the earth's shadow and the sphere at the
time of opposition is represented by the circle ABC. OD is
perpendicular to DM, and since the latter makes with the
ecliptic an angle not greater than five degrees, the length
of OD is a fair approximation to the moon's latitude at the
time of conjunction. If the moon is tangent to the shadow
circle at N, then, in the right triangle ODM, OM is the sum
of the moon's apparent radius and the shadow radius, and DM
is an approximation to the difference between the moon's elongation at first contact and at opposition. DM cannot be
regarded as the difference in lunar longitudes between these

224

times, for in the same period the sun itself, hence the earth's
shadow, will have made some progress along the ecliptic.

When an eclipse is partial, as in the one illustrated in
Figure 16, its magnitude is the length of EC measured in
(eclipse) digits, twelfths of the apparent lunar diameter.
Figure 17, on the other hand, shows a total eclipse. Here

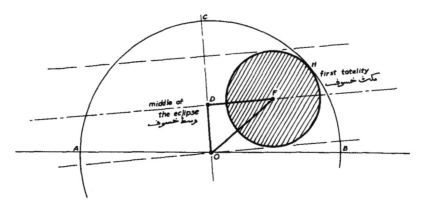

Figure 17. A Diagram for a Total Lunar Eclipse

the beginning of totality is of interest, and this occurs when
the moon is tangent internally to the shadow circle, as at H.
In both cases the approximate middle of the eclipse is marked
by the arrival of the moon's center at D, this corresponding to
the time of opposition. The ends of totality and of contact
are located on the eastward side of OC symmetrically with their
beginnings.

35. Lunar Eclipse Limits

With this background we now examine the text of Chapter

II,11, beginning with f. 31v:1, a statement to the effect that
a necessary and sufficient condition that an eclipse occur is
that the lunar latitude at opposition (β) be less than the sum
of the apparent radii of moon (r_m) and shadow (r_s). The next
sentence, f. 31v:5, looks corrupt, and comparison with the
corresponding passage in NS (p.280) confirms our suspicion.
It says:

> If the lunar latitude at the conjunc-
> tion is more than sixty-three minutes, (then)
> undoubtedly its distance from the node will
> be more than twelve degrees, and so there
> will be no eclipse. But if it is less than
> that and greater than twenty-nine minutes,
> then part of it will be eclipsed.

Although not explicitly stated, it is clear from the figures
that a necessary and sufficient condition that totality be
attained is that

$$r_s - r_m > \beta.$$

The numbers in the Nuzhah lead to the equations

$$r_s + r_m = 63'$$
$$r_s - r_m = 29',$$

whose solution is $r_s = 46'$ and $r_m = 17'$, both numbers being
rounded-off Ptolemaic parameters ([43], ed. of Halma, vol.i,
p.395).

That the latitude condition for a partial eclipse is equivalent to the statement made in our text about the ecliptic
distance from node to opposition may be demonstrated as follows.

COMMENTARY

Consider the right spherical triangle (Figure 18) formed on the
celestial sphere by the node, the moon's center, and the projec-

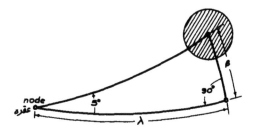

Figure 18. The Lunar Latitude Near Syzygy

tion of the latter on the ecliptic. The relation

$$\sin \lambda \approx \frac{\beta}{5°}$$

subsists, since β and $5°$ are both small. Putting $\beta = 63'$ we
obtain

$$\lambda = \text{arc sin } (63/300) \approx 12°,$$

as demanded by the text.

We are now in a position to investigate the condition with
which Chapter II,11 opens, namely that if the opposition occurs
during daylight it must be less than $2\frac{4}{}$ hours after sunrise or
before sunset. Violation of the condition implies that there
will be no possibility of any part of even the longest eclipse
taking place while the sun is below the horizon.

An eclipse of maximum duration is one for which first contact
takes place when the elongation is sixty-three minutes of arc.

227

Since the mean rate of elongation is about 13;10,15° - 0;59,8°
= 12;11,27° per day, half the length of the longest eclipse is
about

$$\frac{1;3(24)}{12;11,27} = 2;4 \text{ hours.}$$

This is doubtless the origin of the number the author
gives. It does not appear in the Nuzhah.

36. Lunar Eclipse Markings on the Ruler

A permanent mark (O in Figure 19) is placed on the ruler's
edge at a distance of sixty-three parts (sixtieths of the plate
radius) from the end. The mark of first totality (O") is put
permanently on the ruler at a distance twenty-three parts from
the same end. The segment OO" will be thirty-four parts in
length, equal to the apparent diameter of the moon in minutes.
It is divided into twelve equal segments, each one an eclipse
digit.

37. Determination of the Lunar Eclipse Times

Suppose now that β for the time (t) of the opposition has
been computed and is sufficiently small to indicate an eclipse.
Set the alidade point of Aries on the ring. Put a mark O on
the plate such that the segment DO is of length β measured in
divisions of the alidade, taking a sixtieth of the plate radius
for each minute of arc. Rotate the alidade through an angle
of ninety degrees, and place the ruler so that the lunar eclipse
mark on its edge coincides with O, and so that its end touches
the edge of the alidade (at M). The triangle DOM is now a
mechanical construction of the triangle having the same letters
in Figure 16. Hence the length of DM on the instrument gives

228

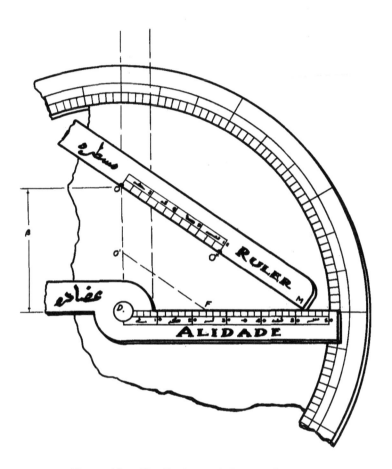

Figure 19.   The Instrument Set Up for a
Lunar Eclipse Determination

the elongation in minutes of arc at the instant of first (or
last) contact. Since the rate of elongation is, very crudely,
a half degree per hour, "depressing" (cf. Section 3) the length

229

of DM once and doubling the result (i.e. dividing by a half)
will yield the elapsed time from first contact to the middle of
the eclipse. This is what the text prescribes. Add the result
to t and subtract it from t to get the times of day at which
last and first contact take place respectively.

If the eclipse is total, an exactly analogous use is made
of the mark of first totality (O") instead of the lunar eclipse
mark (O) to construct the triangle O'FD (in Figure 19) congruent
with OFD (in Figure 17). From it is obtained DM, the elongation
at first totality, which also is converted into time by doubling
and depressing. Addition to and subtraction of the result from
t gives the times of day of last and first totality respectively.

For a partial eclipse the magnitude is measured very simply
by sliding the end (M) of the ruler along the alidade until the
ruler occupies the position indicated by the pair of dotted
lines on Figure 19. The position of the mark O with respect to
the digit scale on the ruler then gives the magnitude in digits
directly. The validity of this procedure can be seen from
Figure 16, where it will be observed that the eclipse magnitude
is, approximately

$$EC = ED + DC = r_m + r_s - \beta = 63' - \beta,$$

which is the construction prescribed.

38. Parallax in the Altitude Circle (f. 35r:4 - 36r:5)

It is convenient here to break the order of the text and
to insert appropriate comment on Chapter II,14, the substance
of which is a preliminary to Section 39.

In Figure 20 assume the small circle with center C to be
the terrestrial sphere. An observer at E takes the altitude of

a celestial body at S, the latter being at a distance SC from
the center of the earth. With respect to the observer's

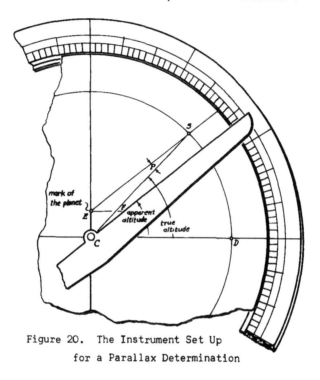

Figure 20.   The Instrument Set Up
for a Parallax Determination

horizon, the tangent to the sphere at E, the altitude of S is
angle SEF. As reckoned from the center of the earth, however,
the altitude is SCD. The difference between these two angles
is P, the parallax in the altitude circle. It is clear that P
is a function of two variables, the earth-planet distance, and
the altitude. For CS constant, P is maximum when the altitude
is zero, taking a maximum value of, very nearly

231

$$\text{arc sin } \frac{EC}{CD}.$$

The operation described in Chapter II,14 consists essentially of making a scale representation of the situation set forth above, and our Figure 20 is sufficient supplement to the author's explanation.

It is of interest, however, to put down his parameters. For the three objects mentioned, horizontal parallax at mean distance will be

☾	arc sin 0;1,2 = 0;59,13°
☉	arc sin 0;0,2 = 0; 1,55°
♀	arc sin 0;0,4 = 0; 3,50°,

where mean distance has been taken as 1,0;0 for the sun and moon and as 1,0;0/2 for Venus, as prescribed in f. 35r:13.

Comparison of these numbers with corresponding entries in a table of parallaxes on f. 185v of Kāshī's zīj [23] show as close a correlation as can be expected. Horizontal lunar parallax at mean distance is there given as 0;59,59°. For the sun it is 0;2,15°, and this accords with our result of 0;1,55°, since the author says vaguely (f. 35r:9) that for the sun CE should be "a little more" than the two minutes we have taken. For Venus the zīj gives 0;4,8°, likewise reasonably close to our value.

For both the sun and Venus the constructions prescribed bear little relation to practical reality. For one thing, the measurement of angles as small as these on an instrument of this sort is out of the question. And for another, the distance of Venus from the earth is based on no observations at all, but on the assumption that the planetary system has been put together

by nesting the successive Ptolemaic configurations as closely
inside each other as will insure no planet's striking its
neighbor.

39. The Table of Parallax Components (f. 33v)

In order that a solar eclipse be visible to an observer it
is necessary that the sun and moon have, for the time of the
eclipse, very nearly the same celestial coordinates. And since
the observer is on the earth's surface, not at its center, it
is clear that the coordinates used must be apparent ones and
not true coordinates, that is, computed with respect to the
observer rather than with respect to the center of the earth.
In other words, it is necessary that a correction for parallax
be applied to the true coordinates resulting from ordinary
determinations.

The adjusted lunar parallax (ikhtilāf-i manzar-i muʿaddal-i
qamar) is the difference between the parallax of the sun and
moon at a time when both have the same apparent altitude. Our
table breaks up this adjusted parallax in the altitude circle
into its latitude and longitude components. Each pair of such
components is for a conjunction occurring at an integer number
of hours before, after, or at the local meridian, and at the
initial point of one of the twelve zodiacal signs. Symmetry
considerations permit the use of a single column for each pair
of signs equidistant from the solstices; thus the second column
from the left (on the transcription) serves for both Leo and
Gemini, provided that the hours for the former are read from
the column at the extreme right, and those for the latter from
the left. The latitude components are given in minutes of arc;
the longitude entries in minutes of time by which the effect of
the parallax will delay or advance the time of apparent conjunction.

233

Since the rate of elongation is about twelve degrees per day, division of the longitude components by two gives an approximate conversion into minutes of arc.

The top row of entries in the table gives the length of daylight when the sun is in the beginning of the sign named at the head of each respective column.

Our author's statement (f. 14r:7) that the table is from Kāshī's zīj is confirmed by examination of the latter document. It is on f. 164v of [23], where the statement is made that the data have been computed for a place of terrestrial latitude 30°. In a special study [27] it has been shown that the numbers in the table are only crudely correct, and that they were probably lifted by Kāshī from some source unknown to us.

40. Solar Eclipses (f. 33r:5 - 34v:3)

The technique for computing solar eclipses with the aid of the instrument closely resembles that used for lunar eclipses, and can be described with reference to what has preceded. Of course it is now conjunctions, not oppositions, which present possibilities for solar eclipses. Just as before, a permanent mark is put on the edge of the ruler, this time at a distance from the end of thirty-three sixtieths of the plate radius. (Cf. f. 12v:7). This implies that the (apparent) lunar latitude at conjunction must be less than thirty-three minutes, a condition which the text gives explicitly in f. 34r:9. The same limit is given in the Almagest ([43] vi,5). The condition also implies that

$$r_m + r_\sigma = 33'$$

where r_σ denotes the apparent radius of the solar disk. Since

234

we have already taken r_m to be about seventeen minutes, it
follows that the solar and lunar disks are assumed to be of
about the same size. Hence even if a solar eclipse is total,
the duration of totality will be very short, and in fact no
mark of first totality is prescribed. The portion of the ruler
edge from the solar eclipse mark to the end is to be divided
into twelve equal parts for the solar eclipse digits (f. 13r:2).

One complication which does not enter in the case of a
lunar eclipse is the fact that for a solar eclipse the geogra-
phical location of the observer is involved. Hence the true
coordinates are to be corrected for parallax by use of the
table on f. 33v. The effect of parallax being always to depress
the apparent position beneath the true one, it follows that if
the conjunction occurs in the forenoon the (longitudinal) correction will be subtracted from the time of true conjunction,
whereas it is added in the case of an afternoon conjunction.
For the same reason the latitude correction is subtracted
algebraically from the true latitude, north being taken as
positive. Once the corrections have been made, the time from
first contact to the middle of the eclipse and the magnitude of
the eclipse are computed on the instrument just as they are for
a lunar eclipse.

The parallax also acts to asymmetrize the necessary condition for a solar eclipse. Its effect is always to pull a celestial object down along a vertical circle from its true position.
Hence when the true moon is north of the ecliptic, the allowable
distance between node and conjunction is greater than when it
is south. Thus the chapter on solar eclipses in the Nuzha
(NS, p.281) begins with the statement that if the conjunction
is after the ascending node and before the descending node,
i.e. if the moon has north latitude, the critical distance is

235

sixteen degrees; if the moon has south latitude the distance
is seven degrees.

The corresponding passage in our text, f. 33r:7-12, doubtless said the same thing in the original composition. However,
it was garbled by the copyist who wrote twice the line and a
half enclosed in braces in the translation. His mistake was
made easier by the fact that the words for distance (bu'd) and
after (ba'd) are written the same unless diacritical marks are
used.

In his zīj ([23], f. 85r) Kāshī gives the same necessary
condition as is found in our text and in the Nuzhah. He
qualifies it by saying that it applies to localities in the
third and fourth climates. A criterion which holds for all
inhabited localities is, he says, that if the lunar latitude
is north, the distance from node to conjunction shall be less
than eighteen degrees; if south the distance shall be less
than nine degrees. He does not derive this condition in the
zīj, and it may very well be rounded off from Ptolemy's 17;41°
and 8;22° arrived at in the Almagest ([43], vi,5).

41. The Solar Mean Longitude at Equinox (f. 34v:4 - 35r:3)

In Islamic astrology the instant at which the sun crosses
the vernal equinoctial point is called the year-transfer
(Arabic tahwīl al-sinah, or simply tahwīl, cf. [4], p.150) and
was considered to be of great significance. Even to the present
in Iran the situation of the individual at the moment of transfer
is supposed to affect his destiny throughout the coming year.

Chapter II,13 in our text explains a method for determining
the time of transfer by use of the equatorium. The same problem
is treated much more elaborately in the Khāqānī Zīj ([23],
ff. 90r - 91r). It can be formulated as the inverse of a more

236

COMMENTARY

common problem, the determination of a planetary true longitude
(λ) as a function of its mean longitude ($\bar{\lambda}$), the latter being
a linear function of time. Now we have the solar true longitude
given, at the equinoctial point ($\lambda = 0°$), and we seek the
corresponding $\bar{\lambda}$. Once this is determined the corresponding
time can be inferred from the mean motion table.

The solutions in the zīj are set up in terms of the mean
and true centers, $\bar{\lambda}_a$ and λ_a, but the difference is trivial, since
addition of the apsidal longitude to a center converts the latter
into a longitude. (Cf. p.188.)

Two apparently alternative methods are given in the zīj
for obtaining $\bar{\lambda}_a$ from λ_a. The first says, find from the table
of solar equations (e) the equation whose argument is the given
λ_a. Add the result to λ_a to obtain a first approximation to $\bar{\lambda}_a$,
and repeat the process. That is, put

$$\bar{\lambda}_{a1} = e(\lambda_a) + \lambda_a,$$
$$\bar{\lambda}_{a2} = e(\bar{\lambda}_{a1}) + \lambda_a,$$
$$\bar{\lambda}_{a3} = e(\bar{\lambda}_{a2}) + \lambda_a.$$

Then, says Kāshī, $\bar{\lambda}_{a3}$ is the desired $\bar{\lambda}_a$. This iterative
method for inverting suitable types of functions is very old,
and probably of Hindu origin (cf. [1] and [32]).

The second method consists of putting

$$\text{Sin } e(\lambda_a) = \sin e_{max} \cdot \text{Sin } \lambda_a,$$

where $e_{max} = 2\text{;}0,29°$ is Kāshī's maximum solar equation. Then

(24) $\bar{\lambda}_a = e(\lambda_a) + \lambda_a = \text{arc Sin}(\sin e_{max} \cdot \text{Sin } \lambda_a) + \lambda_a.$

237

In another part of the zīj, on f. 166v, is a numerical table of the function $L(\lambda_a)$, say,

(25) $$L(\lambda_a) = 2;0,29° \sin \lambda_a + \lambda_a.$$

e_{max} being a small angle, expressions (24) and (25) are approximately equivalent. Thus the table gives a third method for obtaining $\bar{\lambda}_a$.

By contrast with these elaborate and approximate techniques, the solution with the instrument is direct and, at least theoretically, precise. One puts the ruler alongside the fictitious center and parallel to the line joining the plate center and Aries 0°. The point of intersection of the ruler edge with the graduations of the ring gives immediately the mean longitude at which $\lambda = 0°$, the configuration being the same as that for determining λ from $\bar{\lambda}$.

The desired $\bar{\lambda}$ having been determined, it is easy to find from the table of mean motions a date such that at the noon of that day the solar mean longitude $\bar{\lambda}_n$ exceeds the equinoctial $\bar{\lambda}$ just found, whereas on the preceding noon the solar mean longitude is exceeded by the equinoctial $\bar{\lambda}$. The expression $(\bar{\lambda}_n - \bar{\lambda})/\dot{\bar{\lambda}}$ then gives the number of hours from the instant of transfer to the later noon, provided that $\dot{\bar{\lambda}}$ is the rate of change of $\bar{\lambda}$ in degrees per hour.

<u>For material on f. 35r:4 - 36r:5 see Section 38 above</u>

42. The Equation of Time (f. 36r:6 - 37r:8)

Chapter II,15 of the text contains no application of the equatorium, and in fact makes no reference to either one of our instruments. The topic it discusses is, however, of at least

theoretical interest in the major problem solvable with the
equatorium, the determination of planetary true positions. The
table of mean motions, on which such determinations are ulti-
mately based, gives mean positions reckoned from some epoch.
Local apparent time, reckoned from successive meridian passages
of the true sun, differs from mean time because of two facts.
These, as the author points out, are the variable angular
velocity of the true sun in the ecliptic, and the unequal
projections on the celestial equator of equal segments on the
ecliptic. The difference is the equation of time.

The first part of the chapter is copied verbatim from
the extensive equation of time material in the Khāqānī Zīj
([23], f. 93r). At about f. 36v:10 our author abandons this
source, and from there on leans heavily on Appendix 8 (p.311)
of NS. The NK version has nothing on the equation of time;
the subject is another afterthought of Kāshī, but his pre-
sentation is much more satisfactory than that of our text,
which slavishly copies his mistakes.

For instance, f. 37r:3-37r:6 in the text is on the conversion from degrees of arc to time, 360° being equivalent to
twenty-four hours. A degree of arc does correspond to four
minutes of time, and a minute of arc to four seconds, but ten
minutes of arc corresponds to forty seconds of time, not to a
minute, as both the NS and our text have it.

In order that this chapter be useful the author should
have provided the user with specific means of determining the
equation of time. One way would be to give a numerical table
such as that in Kāshī's own Khāqānī Zīj ([23], ff. 126v, 127r).
Another way would be to give explicit directions for computing
both components of the equation of time. The determination of
one component, the solar equation, has already been described,

239

in Chapter II,6. Nothing has been done in the text, however, about the computation of the right ascensions which make up the second component. In NS (p.311) Kāshī describes a technique with the instrument for solving this spherical trigonometric problem. The methods resemble those used for the determination of the lunar latitude and commented on in Section 25 above.

To clarify the passage f. 36v:11 - 37r:2 we remark that the equation of time (E) at any instant t is the difference between the change of mean solar longitude $(\bar{\lambda})$ from epoch (t_0) to time t, and the change in right ascension (α) of the true sun from t_0 to t. Symbolically this is

$$E(t) = \Delta\bar{\lambda}(t) - \Delta\alpha(t)$$
$$= [\bar{\lambda}(t) - \bar{\lambda}(t_0)] - (\alpha(t) - \alpha(t_0))$$
$$= \bar{\lambda}(t) - \alpha(t) - \bar{\lambda}(t_0) + \alpha(t_0).$$

The rule in the text says

$$E(t) = (\bar{\lambda}(t) + 3;57,30°) - \alpha(t).$$

Comparison of the two expressions shows that we must have

$$\alpha(t_0) - \bar{\lambda}(t_0) = 3;57,30°,$$

that is, the constant to be added to the mean longitude is a number which depends on the epoch of the tables and on the base longitude for which they have been computed. This 3;57,30°, however, has been taken over from NS without change, whereas the mean motion tables of the text have both a different epoch and a different base longitude than those of NS.

240

43. The Plate of Conjunctions (f. 37r:9 - 38r:11)

The construction of the second instrument is easily under-
stood from the description on ff. 15r-17r, especially when
considered in conjunction with the drawing on f. 17v, its
modern counterpart on the opposite page, and Figure 21, reproduced from page 287 of NS.

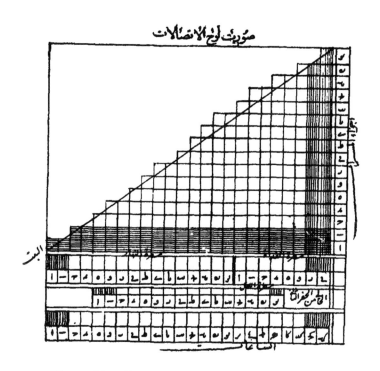

Figure 21. The Plate of Conjunctions, from NS

Of the metrological units mentioned in this passage, the
size of the cubit has been discussed in Section 2 above. The

digit is usually taken as one twenty-fourth of a cubit.

As for the operation of the instrument, the device has been designed to predict the time of day at which a conjunction between two planets will occur, given the noon longitudes of the planets for the given day and for the following day. It is further assumed, of course, that their longitudes show one planet in the lead at the first noon, and the other at the second. Regard the variations in longitude as linear throughout the course of the day. Then if t is the time in hours from the first noon until the conjunction takes place,

$$t = \frac{24d}{b} \text{ ,}$$

where d is the difference between the longitudes of the two planets (Kāshī's past distance), and b is the difference between the daily rates of the two planets (Kāshī's daily motion) for the day in question. Incidentally, the Persian-Arabic term buht, standard for the longitudinal speed of a planet, is from the Sanscrit word bhukti (cf. [4], p.105).

The turning ruler (f. 17r:12, see also Figure 2) is set so that its edge crosses the "divisions of travel" scale at distance b from the beginning of the scale. Find the point on the same scale corresponding to d, from there project horizontally to the edge of the turning ruler, thence vertically down to the base of the triangle. It is clear that the resulting point is at the required distance t from the left vertex of the triangle.

The three sliding scales at the bottom are to convert t from time measured from noon to time measured from local sunrise or sunset, depending on whether the event occurs during the day or the night respectively. To do this, set the day ruler so that the pivot on the turning ruler is opposite the point on the day ruler corresponding to half the length of daylight for the date and locality in question. Set the head of the night ruler opposite the point on the day ruler corresponding to the number of hours of daylight. Put the head of the next-day ruler opposite the point on the night ruler corresponding to the number of hours of darkness. Then, as the author remarks, the right angle of the triangle will fall opposite the point of the next-day scale marking the hour of noon. Now the place where the vertical line distant t units from the pivot intersects one of the three slides at the base gives directly the desired hour of day or night.

In all of the late medieval Persian zījes inspected by the editor, many pages are given over to a double-entry table of the function defining t above (cf. [29], p.162). It is evident that Kāshī invented this simple device to obviate the need for such tables. As such it fulfilled a practical purpose, yielding results of sufficient precision for the problem at hand. The instrument's most serious drawback follows from the fact that the usual daily motion of the planets is of the order of a degree. That of the moon is much larger, averaging over thirteen degrees. This implies that if the conjunction does not involve the moon, the turning ruler would be elevated by so small an angle that the result would be considerably affected by small inaccuracies in the construction.

P E R S I A N   G L O S S A R Y

(Readers who desire to locate in the text one of the words listed below can do so by looking up its English equivalent in the index.)

ا

instrument	آلت
conjunction(s)	اتصالی (pl.) اتصالات
see: ثقب	انقاب
conjunction	اجتماع
see: جزء	اجزاء
difference	اختلاف
parallax	اختلاف منظر
altitude	ارتفاع
difference marks	ارقام اختلاف
forward (motion)	استقامت
direct (station)	(مقام) استقامت
opposition	استقبال
astrolabe	اسطرلاب
digits	اصابع
see: قسم	اقسام
climate(s)	اقلیم (pl.) اقالیم

see: لون الوان

clearance (of an eclipse)	انجلار
obliquity	انحراف
solstices	انقلابین
elliptical (?)	اهلیلجی
apogee(s)	اوج (pl.) اوجات
mean (motions and positions) see: وسط	اوساط

ب

substitute	بدل
sign(s) (zodiacal)	برج (pl.) بروج
slow	بطی
distance(s)	بعد (pl.) ابعاد
doubled distance	بعد مضاعف
rate	بهت

پ

compass	پرگار

245

south	جنوب		ت
node, lunar	جوزهر	calendar or date	تاریخ
sine	جيب	complete (of years)	تامّ
ح		transfer	تحویل
acute (angle)	حادّه	epicycle	تدویر
product	حاصل ضرب	doubling	تضعیف
ring	حجرة	equation	تعدیل
motion(s) (pl.) حكات حركت	equation of time	تعدیل الایام بلیالیها
computation	حساب	differences	تفاضل
argument	حصه	subtraction	تفریق
perigee	حضیض	intersection	تقاطع
depression, or trough	حفر	true (celestial) longitude	تقویم
true	حقیقی	halving	تنصیف
ring	حلقه	succession of the signs	توالی بروج
Aries	حمل		
خ			ث
quotient	خارج قسمت	hole, drill hole	ثقب
eccentric	خارج مركز		ج
anomaly compound anomaly	خاصه خاصه مركبه	table(s) (pl.) جداول جدول	
lunar eclipse	خسوف	part(s), or division(s) (pl.) اجزاء جزء	
wood	خشب	addition	جمع

246

ف

angle	زاویه
right angle	زاویه قائمه
increasing	زاید
projection, excess, lug	زایده
tongue	زبانه
Saturn	زحل
Venus	زهره
astronomical handbook	زیج

س

year	سال
complete (or elapsed) year	سال تامه
speedy, fast	سریع
plane	سطح مستوی
the two inferior planets, Mercury and Venus	سفلین
immersion (of an eclipse)	سقوط
chain	سلسله
year(s)	سنه (pl.) سنین
hole	سوراخ
movement	سیر

line(s)	خطّ (pl.) خطوط
line of apsides	خط اوجی
latitude lines	خطوط عرض
thread	خیط

د

circle(s)	دایره (pl.) دوایر
latitude circle	دایره عرض
degree(s)	درجه (pl.) درجات
rotation, or revolution	دور

ذ

cubit	ذراع
epicyclic apogee	ذروه

ر

retrograde	راجع
quadrant(s)	ربع (pl.) ارباع
retrogradation, or retrograde	رجعت
elevate (a sexagesimal), to	رفع کردن
numeral(s), or mark(s)	رقم (pl.) ارقام
difference mark	رقم اختلاف
string	ریسمان

247

alidade	عضاده	**ش**	
Mercury	عطارد	plumbline	شاقول
magnitude (of a star)	عظم	day, nychthemeron	شبانه روز
node	عقده	yellow copper	شبه
mark	علامت	sun	شمس
mathematics	علم رياضى	**ص**	
astronomy	علم نجوم	brass	صفر
superior (planets)	علويه	plate	صفيحه
غ		**ض**	
extremity, extreme (value), maximum	غايت	multiplication	ضرب
ف		**ط**	
Persian	فارسى	plate of the heavens	طبق المناطق
excess, <u>or</u> difference	فضل	longitude, usually terrestrial longitude, but in the text (f.3r:5) celestial longitude.	طول
ecliptic, <u>or</u> zodiac	فلك البروج	longitudinal	طولى
ق		**ظ**	
base	قاعده	shadow	ظل
perpendicular	قايم	**ع**	
<u>qibla</u>, the direction of Mecca	قبله	universe	عالم
disk	قرص	latitude(s) (pl.) عروض	عرض
Constantinople	قسطنطنيه	bride, (theorem of the)	العروس

248

sine (astrolabe or quadrant) — بجيــب

preserved — محفوظ

axis — محور

circumference — محيط

orbit, on f.8r:11, but in an astronomical context the word usually denotes any circle of the celestial sphere whose pole is the north pole. — مدار

square — ربع

place, the place of a digit in a place-value representation of a number. — مرتبة

elevated — مرفوع

compound (<u>in</u> compound anomaly). — مركبة

center(s), مراكز (pl.) sometimes mean longitude measured from deferent apogee. — مركز

turning center مركز مستعار fictitious center — مركز مدير

apparent — مرئي

pointer — مري

Mars — مريخ

fictitious — مستعار

direct (motion), of a planet — مستقيم

division(s), اقسام (pl.) قسم or part(s)

diameter — قطر

equating diameter — قطر استواء

moon — قمر

pivot, pole — قطب

arc(s) قسي (pl.) قوس

ك

fractions — كسور

solar eclipse — كسوف

star(s), <u>or</u> كواكب (pl.) كوكب planet(s)

total (of an eclipse) — كلى

ل

sights — لبنتين

plate — لوح

color(s) الوان (pl.) لون

م

inclining, <u>or</u> inclined — مايل

explicit (years) — مبسوطة

triangle — مثلث

sum — مجموع

249

Libra	ميزان
inclination, or declination	ميل

ن

decreasing	ناقص
incomplete (of years), or current, or explicit	ناقصه
copper	نحاس
ratio	نسبت
noon, or meridian	نصف النهار
semicircle	نصف دايره
sector(s)	(pl.) نطاقات نطاق
opposite	نظير
decrease	نقص
point(s)	(pl.) نقاط نقطه
latitude point	نقطه عرض
opposite point	نقطه محاذات
light	نور
luminaries, the two, (the sun and the moon)	نيرين

و

chord, or hypotenuse	وتر
mean (position or motion)	(pl.) اوساط وسط

ى

Yazdigerd	يزدجرد

ruler	سطره
peg	مسمار
path(s)	(pl.) مسيرات مسير
Jupiter	مشتري
ascensions, or risings	مطالع
adjusted	معدّل
equant	معدّل المسير (or) معدّل للمسير
equator, celestial	معدل النهار
divisions	مقاسم
station(s)	(pl.) مقامات مقام
true position	مقوم
stationary, stance	مقيم
duration (of an eclipse), first totality	مكث
tangent	مماس
parecliptic	ممثّل
see:	مناطق
depress (a sexagesimal), to	تحط كردن
heaven(s), deferent(s)	(pl.) مناطق منطقه
parallel	موازى
position	موضع

250

BIBLIOGRAPHY

1. Aaboe, A., Al-Kāshī's Iteration Method..., Scripta Mathema-
      tica, vol.20(1954), pp.24-29.
2. Barthold, Wilhelm, Ulugbeg i ego vremya, Zapiski Rossiiskoi
      Ak. Nauk po Ist.-Filolog. Otdeleniyu, VIII Ser.,
      Tom xiii, 1918. German transl. by Hinz, W.,
      Ulug Beg und seine Zeit, Abhandlungen für die
      Kunde des Morgenlandes. D.M.G., XXI Bd., Nr. 1.,
      Leipzig, 1935; see also a note by the same author in
      Izvestiya Imperatorskoi Akademii Nauk, 1914, p.459.
3. Al-Battānī sive Albatenii Opus Astronomicum, edited and
      translated by C.A. Nallino, 3 vols., Milan,
      1899-1907.
4. Al-Bīrūnī, The Elements of Astrology by al-Bīrūnī, edited
      and translated by R. Ramsay Wright, London, 1934.
5. Al-Bīrūnī, Al-Qānun'l-Masʿūdī, 3 vols., Osmania Oriental
      Publications Bureau, Hyderabad-Dn., India, 1954-56.
6. British Museum Catalogue, Catalogus Codicum Orientalium
      Musei Brittannici, Pars Secunda, Codices Arabicos
      Amplectens, Londini, MDCCCLII.
7. Dakhel, A., The Extraction of the n-th Root in the Sexage-
      simal Notation, Unpublished thesis of the American
      University of Beirut, 1951.
8. Dreyer, J.L.E., History of the Planetary Systems from Thales
      to Kepler, Cambridge, 1906; reprinted under the
      title, A History of Astronomy, New York, 1953.
9. Franco, S.G., Catálogo Critico de Astrolabios Existentes
      en España, Madrid, 1945.
10. Ginzel, F.K., Handbuch der mathematischen und technischen
      Chronologie, vol.i, Leipzig, 1906.
11. Gunther, R.T., Early Science in Oxford, vol.ii, Oxford, 1923.
12. Hinz, W., Iran's Aufsteig zum Nationalstaat im fünfzehnten
      Jahrhundert, Berlin, 1936.

13. Hinz, W., Islamische Masse und Gewichte, Handbuch der Orientalistik, Ergänzungsband 1, Heft 1, Leiden, 1955.

14. Honigman, E., Die Sieben Klimata, Heidelberg, 1929.

15. Irani, R.A.K., Arabic Numeral Forms, Centaurus, vol.4(1955), pp.1-12.

16. Kary-Niyazov, T.N., Astronomicheskaya Shkola Ulugbeka, Izdatel'stvo Akademii Nauk SSSR, Moscow, 1950.

17. Al-Kāshī, Jamshīd Ghiyāth al-Dīn, Miftāh al-Hisāb. Published with Russian translation and commentary in [47] below.

18. Al-Kāshī, Nuzhat al-Hadā'iq, India Office MS. 210 in [48], and referred to in this book as NK.

19. Al-Kāshī, Nuzhat al-Hadā'iq, the Samarqand rescension referred to in this book as NS, lithographed in Tehran in 1889 in a single volume with the Miftāh [17].

20. Al-Kāshī, Risālah dar sharh-i ālat-i rasad. The only extant copy is bound with Leyden Cod. 945 (Warner).

21. Al-Kāshī, Al-Risālat al-muhītīyah. Published versions are [36] and [47].

22. Al-Kāshī, Sullam al-samā' = Al-Risālat al-kamālīyah, extant in many manuscript copies, and in a Tehrān lithograph edition of 1306 A.H.

23. Al-Kāshī, Zīj-i Khāqānī fī takmīl al-zīj al-Īlkhānī. The only extant copies are India Office 430 (Ethé 2232) and Aya Sofya 2692.

24. Kennedy, E.S., Al-Kāshī's "Plate of Conjunctions", Isis, vol.38(1947), pp.56-59.

25. Kennedy, E.S., A Fifteenth-Century Planetary Computer: al-Kāshī's "Tabaq al-Manāteq". I. Motion of

the Sun and Moon in Longitude. II. Longitudes,
Distances, and Equations of the Planets, Isis,
vol.41(1950), pp.180-183; vol.43(1952), pp.42-50.

26. Kennedy, E.S., A Fifteenth-Century Lunar Eclipse Computer,
Scripta Mathematica, vol.17(1951), pp.91-97.

27. Kennedy, E.S., Parallax Theory in Islamic Astronomy, Isis,
vol.47(1956), pp.33-53.

28. Kennedy, E.S., The Sasanian Astronomical Handbook Zīj-i
Shāh and the Astrological Doctrine of Transit,
Journal of the American Oriental Society,
vol.78(1958), pp.246-262.

29. Kennedy, E.S., A Survey of Islamic Astronomical Tables,
Transactions of the American Philosophical
Society, vol.46(1956), pp.123-177.

30. Kennedy, E.S., and Muruwwa, Ahmad, Al-Bīrunī on the Solar
Equation, Journal of Near Eastern Studies,
vol.17(1958), pp.112-121.

31. Kennedy, E.S., and Roberts, Victor, The Planetary Theory
of Ibn al-Shātir, Isis, vol.50(1959), pp.227-235.

32. Kennedy, E.S., and Transue, W.R., A Medieval Iterative
Algorism, American Mathematical Monthly, vol.
LXIII(1956), pp.80-83.

33. Khwāndamīr (Ghiyāth al-Dīn ibn Humām al-Dīn al-Husaynī),
Ta'rīkh habīb al-sayr fī akhbār afrād bashar,
Tehran.

34. Krause, M., Stambuler Handschriften islamischer Mathematiker,
Quellen und Studien zur Geschichte der Mathematik,...,
Bd.III, Abt. B, Berlin, 1936, pp.437-532.

35. Luckey, P., Die Ausziehung der n-ten Wurzel und der binomische
Lehrsatz in der islamischen Mathematik, Math.
Ann., vol.120(1948), pp.217-274.

36. Luckey, P., (ed. and transl.), Der Lehrbrief uber den
      Kreisumfang, Abhandlungen der deutschen Akademie
      der Wissenschaften zu Berlin, Jahrgang 1950 Nr.6,
      Berlin, 1953.

37. Luckey, P., Die Rechenkunst bei Gamšīd b. Mas'ūd al-Kāšī,
      Abhandlungen fur die Kunde des Morgenlandes
      XXXI, 1, Deutsche morgenländische Gesellschaft,
      Wiesbaden, 1951.

38. Meshed Catalogue, Oktā'i, Fihrist-i kutub-i kitābkhāneh-i
      mubārakeh-i ostān-i quds-i ridavī, 3 vols., 1345
      A.H.

39. Millás Vallicrosa, J., La introducción del cuadrante con
      cursor en Europa, Isis, 17(1932), pp.218-258.

40. Mosul Catalogue, Dā'ūd al-Chalabī al-Mausilī, Kitāb makhtūtāt
      al-Mawsil, Baghdād, 1927.

41. Neugebauer, O., The Exact Sciences in Antiquity, Second
      Edition, Providence, R.I., 1957.

42. Price, Derek J., The Equatorie of the Planetis, Cambridge,
      1955.

43. Ptolemy, Claudius, Syntaxis mathematica, ed. G.L. Heiberg,
      2 vols., Leipzig, 1898-1903; German transl. by
      K. Manitius, 2 vols., Leipzig, 1912-13; ed. with
      French transl. by M. Halma, Paris, 1813, 1816,
      (reprinted Paris: Hermann, 1927).

44. Rajā'i, Nā'ila, The Invention of Decimal Fractions in the
      East and in the West, Unpublished thesis of the
      American University of Beirut, 1951.

45. Rieu, Charles, Catalogue of the Persian Manuscripts in the
      British Museum, London, 1879-83.

46. Roberts, Victor, The Solar and Lunar Theory of Ibn al-Shātir,
      Isis, vol.48(1957), pp.428-432.

BIBLIOGRAPHY

47. Rosenfeld, Segal, and Yushkevich (ed. and transl.), Dzhamshid
    Giyaseddin al-Kashi, Klyuch Arifmetiki, Traktat
    ob Okruzhunosti,... s prilozheniem reproduktsii
    arabskikh rukopisei oboikh traktatov, Moscow,
    1956.

48. Ross, E.D., and Browne, E.G., Catalogue of Two Collections
    of Persian and Arabic Mss. Preserved in the India
    Office Library, London, 1902.

49. Sauvaire, M.H., Materiaux pour servir a l'histoire de la
    numismatique et de la metrologie musulmane.
    Journal Asiatique, series 8, vol.8(1886), pp.479-
    536.

50. Schmalzl, P., Zur Geschichte des Quadranten bei den Arabern
    München, 1929.

51. Suter, H., Die Mathematiker und Astronomen der Araber...,
    Abhandlungen zur Geschichte der mathematischen
    Wissenschaften..., X. Heft, Leipzig, 1900.

52. Tabātabā'ī, M., Jamshīd Ghiyāth al-Dīn Kāshānī, Āmūzesh
    va parvaresh, vol.10(1319 Hijrī Shamsī), No.3,
    pp.1-8; No.4, pp.17-24.

53. Tabātabā'ī, M., Nāmeh-i pesar beh pedar, Āmūzesh va parvaresh,
    vol.10(1319 Hijrī Shamsī), No.3, pp.9-16, 57.

54. Ulugh Beg, Zīj-i Sultānī. Part of this has been published
    as Prolégomènes des Tables astronomiques d'Oloug-
    Beg, text translation and commentary, by L.A.
    Sedillot, Paris, 1839, 1847, 1853.

References to the text and translation give folio and line, separated by a colon. The corresponding numbers appear along the left-hand edge of each page of the translation. Numbers preceded by an asterisk (*) are references to pages of the commentary.

I N D E X

Lightning Source UK Ltd.
Milton Keynes UK
UKHW022208210122
397535UK00004B/239